COOPERATIVE MONITORING IN THE SOUTH CHINA SEA

COOPERATIVE MONITORING IN THE SOUTH CHINA SEA

SATELLITE IMAGERY, CONFIDENCE-BUILDING MEASURES, AND THE SPRATLY ISLANDS DISPUTES

EDITED BY

JOHN C. BAKER AND DAVID G. WIENCEK

Foreword by Richard H. Solomon

PRAEGER

Westport, Connecticut
London

Library of Congress Cataloging-in-Publication Data

Cooperative monitoring in the South China Sea : satellite imagery, confidence-building measures, and the Spratly Islands disputes / edited by John C. Baker and David G. Wiencek ; foreword by Richard H. Solomon.

p. cm.

Includes bibliographical references and index.

ISBN 0-275-97182-1 (alk. paper)

1. Spratly Islands—International status. 2. South China Sea—Remote-sensing images. 3. Confidence and security building measures (International relations)—Technological innovations. 4. Conflict management. I. Baker, John C., 1949– II. Wiencek, David G., 1958–

KZ3881.S68 C66 2002

341.4'2—dc21 2001054587

British Library Cataloguing in Publication Data is available.

Library of Congress Catalog Card Number: 2001054587
ISBN: 0-275-97182-1

First published in 2002

Praeger Publishers, 88 Post Road West, Westport, CT 06881
An imprint of Greenwood Publishing Group, Inc.
www.praeger.com

Printed in the United States of America

The paper used in this book complies with the Permanent Paper Standard issued by the National Information Standards Organization (Z39.48-1984).

10 9 8 7 6 5 4 3 2 1

To our colleague

This book is dedicated to the memory of our late colleague, Bradford L. Thomas, former political geographer with the U.S. Department of State and professor at George Washington University's geography department. His commitment to helping others to understand better the complexities of international territorial disputes, particularly the Spratly Islands, has been an inspiration to the authors and others who had the pleasure of working with him.

To our families

John Baker dedicates the book to his wife, Meg, and his children, Andrew and Eva.

David Wiencek dedicates the book to his wife, Ruth.

Contents

Tables

Figures

Foreword

The onset of the new millennium coincided with a revolution in information technologies that is having profound implications for nearly every segment of human endeavor. Scholars and practitioners alike are weighing the importance of technologies for generating, processing, and communicating information for the economic, social, and national security aspects of states and societies. One area requiring greater intellectual attention is whether the information revolution can create new opportunities for conflict management and diplomacy in order to reduce the risk of armed conflicts among nations. New information technologies, including high-resolution commercial observation satellites, are emerging that are particularly promising instruments for preventive diplomacy and conflict resolution. Such technologies offer an improved means for reducing the chances of miscalculation and misunderstanding by national leaders, who often make crucial decisions involving the risk of armed conflict in the face of great uncertainty and confusion.

This book focuses on a dangerous regional flashpoint—the disputed Spratly Islands in the South China Sea—where information technologies could play an important role in reducing the likelihood of regional conflict. It is specifically concerned with minimizing the conflict potential of the complex jurisdictional and territorial disputes surrounding the Spratly Islands and their associated waters that arise from the conflicting claims of six countries: China, Vietnam, the Philippines, Malaysia, Taiwan, and Brunei. Although major fighting has been avoided to date, the Spratly Islands disputes encourage continuing regional tensions as claimants compete to occupy the remaining small islands and reefs with military and civilian outposts intended to bolster sovereignty claims. These rivals typically use military reconnaissance aircraft and naval patrols to monitor each other's activities in the Spratlys; despite the danger of provoking a forcible response that could escalate into the wider use of military force.

The editors of *Cooperative Monitoring in the South China Sea*, John Baker and David Wiencek, have done an admirable job of bringing together a multidisciplinary team of experts in regional affairs and satellite technology to assess whether a new generation of commercial and civilian observation satellites can mitigate the risk of armed conflict arising from the Spratly Islands disputes. Their study addresses both the technical and political feasibility of creating a cooperative monitoring regime that would use commercial and civilian satellite imagery as a nonprovocative means of monitoring the disputed Spratly locations compared with close-up aerial overflights and naval patrols. Among the book's highlights is a detailed analysis of satellite imagery of a selected Spratly location, Mischief Reef, that demonstrates the strengths and limitations of high-resolution satellite imagery for monitoring important activities occurring at disputed islands and reefs. Both governments and non-state actors (for example, nongovernmental organizations) will enjoy increasing access to timely satellite images as the number of high-resolution commercial and civilian observation satellites continues to increase.

The authors do not confine their analysis to the technical feasibility of using satellite imagery to monitor politically and militarily sensitive developments. They should be commended for taking a broader approach that weighs the political, economic, and security considerations that will influence the willingness of different South China Sea littoral states to participate in a cooperative monitoring regime. The book highlights the frequently overlooked mutual interests of the Spratly Islands claimants and other Southeast Asian countries, such as Indonesia and Singapore, in sharing information on common problems in the South China Sea region. These problems include maritime pollution, natural disasters, and the continuing scourge of piracy in the South China Sea.

The United States Institute of Peace (USIP) has made special efforts to call attention to the risks of regional conflicts arising from the complex territorial disputes in the South China Sea. During the mid-1990s, the Institute convened a South China Sea Meeting Series that focused on the opportunity for preventive diplomacy as a means of managing the territorial and maritime disputes. Special attention was given to the growing tensions over the Spratly Islands.[1] In addition, the Institute has funded a series of research projects and studies since 1991 that have analyzed the nature of the regional rivalry over the South China Sea, as well as the proposed approaches to solving the underlying political problems or mitigating the risks of unintended armed conflict. Indeed, the present volume is an outgrowth of an Institute grant to assess the role that commercial observation satellites could play in reducing the risks of conflict in the South China Sea.[2] More recently, the Institute published a work acknowledging the continuing importance of the "Workshop on Managing Potential Conflicts in the South China Sea," an annual "track two" meeting held under Indonesian sponsorship.[3] These meetings are an important exercise in informal diplomacy. They create the possibility of sustaining an unofficial dialogue among government and non-governmental experts from countries concerned with developments in the South China Sea despite the national disagreements.

The work presented in this book on the potential use of commercial and civilian observation satellites in a conflict management regime is central to the Institute's "Virtual Diplomacy" initiative, which explores how modern information and communication technologies are transforming diplomacy.[4] One of the Institute's main research objectives is to identify promising means for using information technologies for international conflict prevention, management, and resolution. This book offers a good example of how some of the positive features of modern information technologies can be used to help reduce the risks of armed conflicts arising over longstanding regional disputes.

Cooperative Monitoring in the South China Sea offers a new perspective on a regional flashpoint, which has grown so frustratingly familiar that it is tempting for policymakers, experts, and journalists to conclude that the Spratly Islands disputes are unsolvable by diplomatic means and are likely to be an everpresent trigger for a wider regional conflict. Without claiming to have found a solution to the complex sovereignty disagreements that fuel the multinational disputes over the Spratlys, the authors provide some much-needed fresh thinking on ways to harness new information technologies to help the South China Sea littoral states diminish the risk of warfare over the Spratly Islands.

Richard H. Solomon, President,
United States Institute of Peace

NOTES

1. *The South China Sea Dispute: Prospects for Preventive Diplomacy: A Special Report of the United States Institute of Peace* (Washington, D.C.: United States Institute of Peace, August 1996).
2. USIP funding helped to initiate the "South China Sea Remote Sensing Project," which was sponsored and supported by The George Washington University's Space Policy Institute during 1997–1999. Project contributors include John C. Baker (Space Policy Institute), Dr. Adam Bernstein (Sandia National Laboratories), Dr. Vipin Gupta (Sandia National Laboratories), Professor Bradford Thomas (George Washington University's Geography Department), David Wiencek (International Security Group, Inc.), and Dr. Ray A. Williamson (Space Policy Institute).

3. Hasjim Djalal and Ian Townsend-Gault, "Managing Potential Conflicts in the South China Sea: Informal Diplomacy for Conflict Prevention," in *Herding Cats: Multiparty Mediation in a Complex World*, edited by Chester A. Crocker, Fen Osler Hampson, and Pamela Aall (Washington, D.C.: United States Institute of Peace Press, 1999).
4. For additional information, see the descriptions of USIP's continuing work on "Virtual Diplomacy" questions at http://www.usip.org/oc/virtual_dipl.html.

Abbreviations

AFP	Armed Forces of the Philippines
ARF	ASEAN Regional Forum
ASEAN	Association of Southeast Asian Nations
ASG	Abu Sayyaf Group
AVHRR	Advanced Very High Resolution Radiometer
CASI	Compact Airborne Spectrographic Imager
CBERS	China-Brazil Earth Resources Satellites
CBM	Confidence-Building Measure
CCTV	Close Circuit Television
CREWS	Coral Reef Early Warning Systems
CRISP	Centre for Remote Imaging, Sensing and Processing
CSCAP	Council for Security Cooperation in Asia Pacific
CZCS	Coastal Zone Color Scanner
DTM	Digital Terrain Model
EEZ	Exclusive Economic Zone
ERS	Earth Resources Satellites
ESA	European Space Agency
GIS	Geographic Information System
GOES	Geostationary Operational Environmental Satellite
GPS	Global Positioning System
ICC	International Chamber of Commerce
IMB	International Maritime Bureau
ISAR	Inverse Synthetic Aperture Radar
ISRO	Indian Space Research Organization

JERS	Japan Earth Resources Satellite
LLAR	Low Level Armed Robbery
LLTV	Low Light Level Television
LNG	Liquefied Natural Gas
MCHJ	Major Criminal Hijack
MERIS	Medium Resolution Imaging Spectrometer Instrument
MILF	Moro Islamic Liberation Front
MIMA	Malaysian Institute of Maritime Affairs
MLAAR	Medium Level Armed Assault and Robbery
MODIS	Moderate Resolution Imaging Spectrometer
MNLF	Moro National Liberation Front
MSS	(Landsat) Multi-Spectral Scanner
NASA	National Aeronautics and Space Administration
NGO	Nongovernmental Organization
NIR	Near Infrared
NOAA	National Oceanic and Atmospheric Administration
PLA	People's Liberation Army
PRC	People's Republic of China
ROCSAT	Republic of China Satellite
RPC	Regional Piracy Center
SAR	Synthetic Aperture Radar
SCSMEX	South China Sea Monsoon Experiment
SDF	Self Defense Forces
SeaWiFS	Sea-viewing Wide Field of View Sensor
SLOC	Sea Lines of Communication
SPEX	Shell Philippines Exploration
SST	Sea-Surface Temperatures
TDMA	Time Division Multiple Access
TIANS	Thai Institute of Advanced Naval Studies
TM	(Landsat) Thematic Mapper
TOMS	Total Ozone Mapping Spectrometer
TRMM	Tropical Rainfall Measuring Mission
UNCLOS	United Nations Convention on the Law of the Sea

1

Introduction

John C. Baker and David G. Wiencek

A fortuitous result of the emerging information age is that new instruments for supporting conflict prevention and resolution are becoming widely available. This book examines how a new information technology—commercial observation satellites capable of collecting high-resolution images—could be used to mitigate the risk of regional conflict arising from the territorial disputes over the Spratly Islands in the South China Sea. Although information technologies are unlikely to resolve the complex sovereignty disputes over the Spratlys, they do offer diplomats and defense planners a potentially important instrument for reducing the risk that political competition for the Spratlys will inadvertently escalate into armed conflict. Commercial and civilian satellite imagery also could provide a basis for greater cooperation on regional problems, such as monitoring environmental problems or combating maritime piracy, where the littoral states of the South China Sea have common interests despite their other political disagreements.

The persistent territorial disputes over the Spratly Islands pose a challenging problem for assessing whether modern information technologies, particularly high-resolution commercial observation satellites, can play an important role in mitigating the risks of armed conflicts by enhancing regional transparency and promoting regional cooperation. These desolate islands and reefs have gained international notoriety because multiple states (such as China, the Philippines, Vietnam, Malaysia, Taiwan, and Brunei) each claim all or some portion of the Spratly Islands and their adjacent waters. Most of these countries have buttressed their claims by establishing military outposts and other forms of permanent human presence on various islands and reefs. Uncertainty over the intentions and actions of the competing states contributes to diplomatic tensions and sustains the Spratlys as a major flashpoint for regional conflict.

The multinational character of the Spratlys dispute ranks it among the world's most complicated territorial disagreements. The protracted dispute over the Spratlys perpetuates the risk of a more violent clash occurring that could undermine regional stability. These territorial disputes in the heart of the vital maritime passages of the South China Sea therefore present an important case for assessing whether an emerging information technology—commercial satellite imagery—can help defuse regional conflicts by enhancing transparency and promoting greater cooperation.

THE SPRATLY ISLANDS DISPUTES

The Spratly Islands are situated in the southern reaches of the South China Sea (see Figure 1.1). By one count, they consist of some 170 low-lying features. This includes about three dozen small islands along with various reefs and submerged banks and shoals.[1] The total land area of the tiny islands is no more than 2 to 3 square kilometers in an ocean area covering over 200,000 square kilometers. This maritime region's known and potential natural resources only exacerbate the sovereignty disputes. The regional conflicts over the South China Sea arise from competing national interests in exploiting some of the world's richest fishing waters, as well as the hopes of the littoral states that potentially large reserves of oil and natural gas exist within the South China Sea that can be economically exploited.

A growing dispute over the political jurisdiction of the Spratly Islands has arisen over the past three decades. Five of the claimant countries (Taiwan, Vietnam, China, Malaysia, and the Philippines) have established a continuous human presence on different small islands and at some of the key reefs in the Spratly Islands (see Appendix). Their military outposts and other facilities serve to demonstrate the seriousness of their sovereignty claims.[2] This competitive occupation of key features has sporadically grown over the past five decades as one country and then another staked out its territorial and maritime claims to the Spratly Islands.

The initial postwar occupation began in the mid-1950s when Taiwan occupied Itu Aba island, which remains its sole possession among the Spratly Islands. The pace of occupations increased during the 1970s when the Philippines began occupying several Spratly islands. Several islands were initially occupied by South Vietnamese forces in the early 1970s and then replaced by troops sent by Hanoi.[3] However, the number of occupied locations more than tripled during the 1980s as Vietnam, China, Malaysia, and the Philippines all expanded their occupied holdings in the Spratly Islands. Occupation of new locations in the Spratly Islands during the 1990s have been relatively few in number but have substantially increased the level of

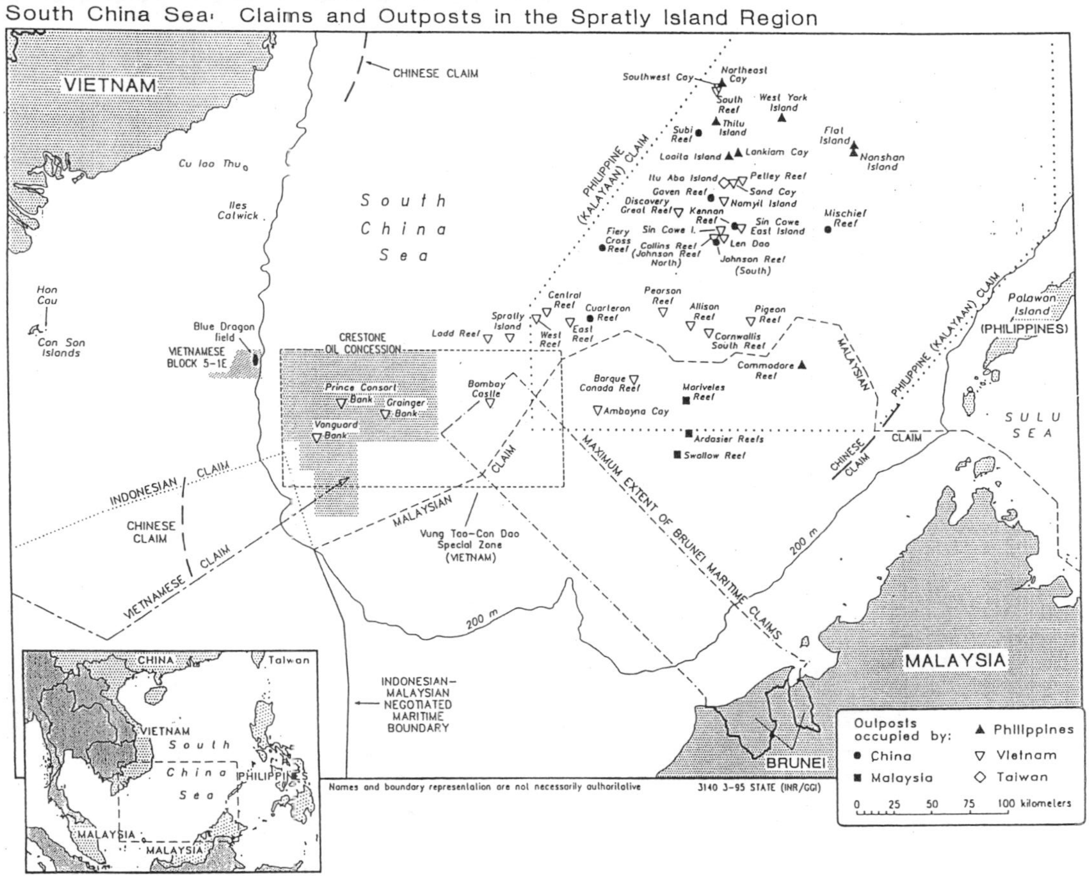

Figure 1.1

South China Sea: Claims and Outposts in the Spratly Island Region.

regional tensions in some cases, such as Chinese activities at Mischief Reef during 1995 and 1998–1999. One estimate (see Table 1.1) is that about 1,500 troops are deployed on some 47 islands and reefs in the Spratly Islands and sustained by the various claimant countries. However, as discussed in Chapter 4, "Security Risks of a South China Sea Conflict," this estimate is outdated given the new occupations by Vietnam and Malaysia, and China's major expansion in the living facilities at Mischief Reef.[4]

Table 1.1
Estimated Occupational Forces in the Spratly Islands[5]

Country	*Number of Islands Occupied and Some Key Features*	*Estimated Number of Troops*
PRC	7 islands and reefs; several helicopter pads	260
Philippines	8 islands; one with a 1,300-meter runway	480
Vietnam	27 islands and reefs; one with a 600-meter runway	600
Malaysia	4 islands; one with a 600-meter runway	70
Taiwan	1 island with helicopter pads; plans for runway	100

Each country offers particular historical or legal rationales to support its jurisdictional and territorial claims to the disputed Spratly Islands.[6] However, at least three common factors underlie their claims and help explain why these countries are willing to expend their limited resources in maintaining a significant presence at these remote locations—sovereignty, fishing resources, and hydrocarbon potential.

SOVEREIGNTY AT STAKE

The sovereignty issue alone offers a powerful impetus for the states engaged in this territorial dispute. Fused with a strong dose of nationalism, it has helped make political-security considerations a dominant factor in the claimants' approach to the dispute.

The competing claimants have populated various Spratlys islands and reefs with small garrisons or other forms of human presence intended to "show the flag."[7] Maintaining a permanent presence at these remote locations is not an easy undertaking for the claimants to the Spratlys because these locations must depend on continuing resupply over long distances from their home countries (for example, China's outposts on Mischief Reef are over 1,000 miles from the Chinese mainland). And despite occasional stories touting the patriotic commitment of their troops in showing the flag among the disputed Spratlys, even the official news accounts reveal that the personnel deployed on these small islands and reefs must endure harsh conditions at times. As one Chinese newspaper account noted, "every bit of soil and food and every drop of fresh water must be delivered from the interior," to support these remote locations and that the soldiers on the reefs must endure "shortages of fresh water and vegetables, loneliness, bad weather and hard life" in performing their mission.[8] Thus, the competing claimants to the Spratlys are making a substantial national commitment by constructing habitable facilities in these remote locations and maintaining a small peacetime presence of troops or other personnel.

Growing nationalism among many Asia Pacific nations following the end of the Cold War only increases the domestic constraints on national leaders in exhibiting diplomatic flexibility in the Spratly Islands disputes. Even if competition for natural resources was not a salient issue, as it is in the South China Sea, sovereignty is therefore a sufficient condition for countries to assume intractable positions and to take actions in the area that increase the risk of armed conflict.

RICH BUT DWINDLING FISHING GROUNDS

The dispute over the Spratly Islands is reinforced by competition for maritime resources, particularly fishing and energy resources.[9] Competition for dwindling fish stocks is a key feature of the Spratly Islands disputes, as well as a general source of maritime conflict among the littoral states of the South China Sea. The scramble for such potentially valuable maritime resources is exacerbated by how some countries interpret the United Nations Convention on the Law of the Sea (UNCLOS) to support their national claims for an Exclusive Economic Zone (EEZ) extending out to 200 nautical miles, based on continuously occupying key locations among the Spratly Islands.[10]

The rich fishing grounds of the South China Sea offer an indispensable source of food and protein for many Asian nations, as well as provide an important source of employment and substantial foreign exchange earnings.[11] The long-term productivity of the South China Sea is declining, however, under the combined pressures of overfishing, coral reef damage, and growing coastal pollution. Hence, the competing national governments are particularly sensitive to incursions by foreign vessels into the rich fishing grounds that they claim for their own use and that they perceive as a dwindling resource.

STRONG INTERESTS IN POTENTIAL ENERGY DEPOSITS

Diplomatic efforts to settle the conflicting claims are hindered by perceptions that the South China Sea could be a possible major source of hydrocarbon resources for the energy-hungry developing countries of the region. According to some studies, the South China Sea is potentially rich in energy resources, such as oil and natural gas. China, Vietnam, and several other claimants have a strong need for reliable and inexpensive energy sources to sustain their future economic development.

The South China Sea has been touted for years as being possibly rich in both oil and natural gas resources. However, despite years of active exploration for oil deposits and limited offshore drilling operations, the extent of the oil reserve, particularly oil that is economically recoverable, remains highly uncertain. In comparison, natural gas could be the more abundant hydrocarbon resource in the South China Sea, and could be more economical to recover. China and Malaysia are already developing their offshore fields for natural gas production, and Indonesia has begun developing the large Natuna gas field located in the southern region of the South China Sea.[12]

Substantial uncertainty surrounds the extent and accessibility of hydrocarbon resources in the South China Sea, and particularly in the vicinity of the Spratly Islands. Nonetheless, the claimant states are quick to view energy exploration and recovery operations by other states as encroachments on their territorial and maritime claims. The presence of oil and natural gas reserves therefore only exacerbates the sovereignty disputes among the various South China Sea claimants.

NATURE OF THE CONFLICT

Although the contending countries have largely engaged in a war of words and diplomatic protests, incidents involving the use of force and even the loss of life periodically occur in the South China Sea. On several occasions, these disputes have escalated into the limited use of force and outright fighting. In 1971, for example, Taiwan allegedly fired on a Philippine vessel attempting to land on Itu Aba. In 1974, China displaced South Vietnamese troops from the Paracel Islands group in the northern area of the South China Sea.[13] And in 1988, China and Vietnam engaged in the most violent clash to date in the Spratly Islands when an incident escalated into armed conflict that resulted in the sinking of some Vietnamese transport ships with most of their sailors killed or missing in action.

During the mid-1990s tensions arose between China and the Philippines over Chinese activities within the territorial waters proclaimed by Manila. In early 1995, China established an outpost at Mischief Reef by building a series of dwellings on stilts upon the reef structure. The Chinese have claimed that these shelters were erected to protect their fishermen against the elements. Manila strongly protested the new Chinese outpost. Manila also received support from the other Association of Southeast Asian Nations (ASEAN) countries that were worried about Chinese moves to buttress its broad claims to the entire South China Sea. Diplomatic tensions subsided until late 1998 when China undertook a major expansion of the facilities at Mischief Reef, once again drawing strong protests from Manila. However, China was not alone in expanding its permanent presence in the Spratlys; by 1999, both Vietnam and Malaysia had reportedly established some new outposts. Thus, regional rivalry over the Spratlys fuels a persistent pattern of territorial and maritime disputes, which is only exacerbated by continuing national disagreements over claims to offshore energy resources and fishing areas. The incident in April 2001, which involved a midair collision between a Chinese interceptor and an American EP-3 reconnaissance aircraft, further highlights the unpredictable risks of conflict arising in the South China Sea.

The latent risk of armed conflict arising over the Spratly Islands is all the more dangerous because of its seemingly routinized nature, which serves to conceal an underlying unpredictability. These territorial disputes are mostly characterized by diplomatic exchanges, as well as by occasional provocative actions, such as fishing ship incursions or oil explorations in the disputed areas by one claimant that stimulates another claimant to undertake a strong diplomatic response or even a show of force. However, the 1988 naval clash between China and Vietnam offers evidence that some incidents in the Spratly Islands have escalated into a more serious showdown, including the use of deadly force.

Unfortunately, armed conflict in the South China Sea involving the claimant countries also will create significant problems for other Asian countries, such as Japan, South Korea, Australia, and Singapore. Fighting over the Spratly Islands could disrupt the critical sea lanes that pass through the South China Sea. Economic development in the region could be hurt by disrupted maritime trade and strained diplomatic relations. And a serious confrontation between China and an ASEAN country, such as the Philippines, would place the United States in a very difficult foreign policy position—torn between a need to support an American ally and not being eager to take actions that would damage U.S.-China political and economic relations.

This cyclical pattern of confrontations and accommodations over the competing South China Sea claims might elicit little concern among non-claimant countries, except that historical precedents exist where long simmering territorial disputes have eventually erupted into substantial armed conflicts with important regional security implications. A good example is the sovereignty dispute between Argentina and the United Kingdom that suddenly escalated into the Falklands War in mid-1982. The fighting over the Falklands, which took the leaders and publics on both sides by surprise, suggests that countries often undertake diplomatic risks and military actions with unintended consequences, including major warfare.[14] Thus, the Falklands War is a prudent reminder that countries will resort to force when they believe that their sovereignty is being unacceptably challenged. Measures to enhance conflict avoidance therefore are essential to reduce the chances that an unexpected confluence of external and internal political factors can produce war.

TRADITIONAL MEANS FOR ENCOURAGING CONFLICT PREVENTION

Diplomacy has been the traditional instrument for attempting to prevent an armed conflict from arising over the Spratly Islands disputes. Both formal institutions and informal or "track two" dialogues are central to diplomatic efforts aimed at regional conflict prevention. However, in the case of the Spratlys, China's disproportionate military and economic power creates a complex challenge for the various ASEAN countries. The smaller countries are anxious to avoid confronting their much more powerful neighbor over the competing claims to the Spratlys. Most of the ASEAN countries also have important trade ties with China, which they desire to expand. Nonetheless, the Philippines, Malaysia, and Vietnam have shown no willingness to concede their sovereignty claims in the Spratlys by accepting China's expansive claim to the entire South China Sea.

The ASEAN countries look to multinational venues for dealing with China on the contentious Spratly Islands issues. The ASEAN Regional Forum (ARF), a large multilateral forum for consultations among countries with interests in Asia-Pacific security issues, is the main multinational forum where the ASEAN countries try to engage China on ways to defuse tensions over the Spratly Islands by agreeing to a 'code of conduct' and considering possible confidence-building measures. The potential leverage offered by the ASEAN countries standing together in dealing with China on South China Sea issues was highlighted in the August 1995 meeting of the ARF, which occurred in the wake of the Mischief Reef showdown between China and the Philippines. The meeting was successful in that China pledged to respect the right of safe passage and freedom of navigation through the South China Sea.[15] However, China's subsequent move to expand its outpost at Mischief Reef renewed concerns among many ASEAN countries over Beijing's willingness to honor its earlier pledges. Efforts to create a dialogue on the South China Sea disputes within the ARF agenda have been repeatedly inhibited by Beijing's preference for dealing with these issues on a bilateral basis with the other claimant countries.

Unofficial or "track two" discussions concerning the South China Sea issues, including the Spratlys disputes, have been operating in parallel with the more formal governmental discussions since about 1990.[16] This process has produced a series of technical working groups that focus on issues of common concern to the littoral states of the South China Sea, including scientific research, maritime safety, and environmental protection issues.[17] Another important forum for unofficial discussions is the Council for Security Cooperation in Asia Pacific (CSCAP), which fosters an exchange of views among experts concerned with regional problems, including the South China Sea disputes.

INFORMATION TECHNOLOGIES FOR CONFLICT PREVENTION

Despite continuing official and nongovernment efforts, only limited progress has been achieved in resolving the territorial and maritime disputes among the various claimants to the Spratly Islands. This situation places a premium on identifying new approaches for conflict prevention and for mitigating the risk of conflict arising among the claimant countries. One promising possibility is to draw on improving information technologies, including the growing availability of commercial observation satellites to mitigate the risk of inadvertent armed conflict in the South China Sea.

This book contends that high-resolution commercial satellite imagery offers an increasingly feasible and less provocative means for monitoring politically and militarily significant developments occurring in the Spratly Islands. Particularly provocative developments could include the establishment of new outposts or a major upgrade to existing military outposts/facilities, such as what occurred at Mischief Reef in 1998–1999. Compared with close-to-shore naval patrols or low-altitude aerial overflights, commercial observation satellites can acquire detailed information on these locations and facilities without the risk of provoking a forcible response from one of the Spratlys claimants.

A new generation of commercial observation satellites first became available in mid-1999 with the launch of Space Imaging's *Ikonos* satellite. These new satellites offer much improved image resolution and greater timeliness in collecting and disseminating their imagery data compared with earlier non-military satellites. Equally important, a growing number of commercial (and civilian) observation satellites are projected to become operational over the next few years with the expectation of rapidly expanding global access to most of their imagery data products.[18]

Of course, high-resolution commercial observation satellites will not resolve the underlying sovereignty issues that fuel the multinational disputes over the Spratlys. Yet, they do offer a new instrument for 'virtual diplomacy' by offering diplomats and defense planners a timely source of accurate information that can help reduce the risk of an unintended armed conflict arising over the Spratlys. Commercial and civilian satellite imagery creates novel opportunities for data sharing among the littoral countries of the South China Sea. In addition, it also can help both Spratly Islands claimants and other littoral states deal with important common problems, such as maritime pollution, natural disasters, and maritime piracy.

This book assesses the potential role that a new information technology—high-resolution commercial observation satellites—can play in conflict prevention regarding the Spratly Islands disputes in the South China Sea, a region where seemingly lower-level incidents run a real risk of escalating into wider armed conflict. It is designed to contribute to the literature on the South China Sea disputes and wider Asian political-security studies generally. While the issues relating to the South China Sea disputes have been addressed in many media and journal articles (see Recommended Readings), there are very few book-length treatments of the subject available. In addition, we have tried to capture the most recent thinking on the application of emerging high-resolution commercial satellite imagery as it pertains to confidence building and regional diplomacy. Thus, we hope this book will find an audience among diplomats, scholars, and students, as well as defense planners and security specialists.

As part of our broader assessment of whether commercial and civilian observation satellites can in fact contribute to regional conflict avoidance and resolution, we address the following questions:

- What is the nature of the multinational territorial disputes over the Spratly Islands?
- Can commercial and civilian satellite images detect and identify politically and militarily significant activities being undertaken by the territorial claimants in the Spratlys?
- What remote sensing experience do the countries most concerned with developments in the South China Sea already possess?
- How could a regime that promotes great transparency concerning the Spratly Islands disputes be designed?
- What are the broader political incentives and disincentives for the littoral countries to participate in cooperative monitoring for the South China Sea?

These questions are addressed in the three parts of the book. Part I provides the necessary context for understanding the South China Sea disputes. Without attempting to be exhaustive, several chapters analyze the Spratly Islands disputes against the broader diplomatic, military, and energy resource developments in the South China Sea. In Chapter 2, "The Spratly Islands Dispute: Legal Issues and Prospects for Diplomatic Accommodation," Christopher Joyner reviews the diplomatic and legal disagreements over which of the claimants countries has legitimate claims for sovereignty over the Spratlys. In Chapter 3, "Energy Issues in the South China Sea Region," Erik Kreil offers some much-needed perspective on actual versus perceived energy resources of the South China Sea, which are rightly or wrongly often portrayed as the main source of the competition over the Spratlys. The regional security risks associated with the jockeying for position in the South China Sea among the claimant countries are analyzed in Chapter 4, "Security Risks of a South China Sea Conflict," by David Wiencek and John Baker.

Part II focuses on the potential role of non-military imaging satellites in supporting conflict prevention and resolution in the South China Sea, with particular attention being paid to improving transparency concerning the Spratly Islands disputes. In Chapter 5, "Commercial Observation Satellites for Conflict Avoidance," John Baker assesses the trends in commercial and civilian observation satellites that make them much more capable for supporting security-related applications. In Chapter 6, "Cooperative Remote Monitoring in the South China Sea: Comparing the Utility of Aerial and Satellite Imagery," Vipin Gupta and Adam Bernstein provide a more technical imagery analysis of whether such satellites can provide imagery of sufficient resolution to detect and identify politically significant developments occurring among the Spratly Islands, such as the construction of new facilities or the presence of naval ships. Chapter 7, "Remote Sensing Capabilities and Experience Available to the South China Sea Littoral States," by John Baker draws heavily on the earlier work on this project of our late colleague, Bradford Thomas, to survey the remote sensing experience and capabilities of the key littoral states of the South China Sea.

Part III provides chapters that recognize that the prospects of realizing cooperative monitoring in the South China Sea depend mainly on the interest of both the Spratly claimants and interested non-claimant countries to collaborate on promoting greater regional transparency. Chapter 8, "A Cooperative Monitoring Regime for the South China Sea," by John Baker and David Wiencek offers an illustrative regime for cooperative monitoring of the disputed Spratly Islands that presents a monitoring strategy and identifies the required imagery analysis staff. Chapter 9, "Environmental Remote Sensing of the South China Sea," and Chapter 10, "Contemporary Maritime Piracy in Southeast Asia," highlight regional problems beyond the Spratly Islands disputes where the littoral states share mostly common interests for cooperative monitoring. Douglas Fuller in Chapter 9 examines the contribution that improved civilian and commercial observation satellites can make to monitoring environmental trends and problems in the South China Sea. In comparison, Peter Chalk in Chapter 10 assesses the nature of the maritime piracy that plagues Southeast Asian waters. He also considers how new information technologies, including commercial observation satellites, can help regional governments in combating the piracy problem. In Chapter 11, the concluding chapter, the editors draw together the implications of the earlier chapters for the desirability and feasibility of pursuing a cooperative monitoring regime for the South China Sea, as well as assess the diplomatic acceptability of these arrangements for key countries, including China and the ASEAN countries. Finally, the Appendix presents information on the occupied locations among the Spratly Islands that supplement the main chapters.

NOTES

1. Daniel J. Dzurek, *The Spratly Islands Dispute: Who's On First?*, Maritime Briefing 2 (Durham, U.K.: University of Durham, International Boundaries Research Unit, Department of Geography, 1996), 1–3. Identifying particular locations can be a problem because each of the territorial claimants has its own names for many of the key features among the Spratlys. For example, China and Taiwan call this area the *Nansha* Islands or Archipelago, while Manila calls the portion of the Spratly Islands that it claims *Kalayaan* or "Freedomland." For purposes of convenience, this book uses the traditional English names. For a comparison of the names used by different claimants for the same disputed locations, see Table 2 in Daniel J. Dzurek, *The Spratly Islands Dispute: Who's On First?*, 56–57.
2. Scott Snyder, *The South China Sea Disputes: Prospects for Preventive Diplomacy, A Special Report of the United States Institute of Peace* (Washington, D.C.: United States Institute of Peace, August 1996), 5.
3. Daniel J. Dzurek, *The Spratly Islands Dispute: Who's On First?*, Maritime Briefing 2 (Durham, U.K.: University of Durham, International Boundaries Research Unit, Department of Geography, 1996), 20–21.
4. Vietnam reportedly sent troops to seize two additional submerged reefs in the Spratly Islands (probably Orleana Shoal and Kingston Shoal on Rifleman Bank) in September 1998 and Malaysia constructed new facilities at Investigator Shoal in June 1999. These actions could increase the number of occupied locations in Table 1.1 to about 49 occupied locations in the Spratly Islands. A classified U.S. Government report obtained by the Kyodo news agency placed the number of Chinese troops in the Spratlys in 1999 at 325. See Dario Agnote, "China to Continue Military Buildup in Spratly Islands," *Washington Times*, (October 22, 1999), A14.
5. Cheng-yi Lin, "Security Implications of Conflict in the South China Sea: A Taiwanese Perspective," in Carolina G. Hernandez and Ralph Cossa, eds., *Security Implications of Conflict in the South China Sea: Perspectives from Asia-Pacific*, (Quezon City, the Philippines: Institute for Strategic and Development Studies, Inc., 1997), 104; and Bradford L. Thomas and Daniel J. Dzurek, "The Spratly Islands

Dispute," *Geopolitics and International Boundaries* 1 (Winter 1996), 302–306; and Rigoberto Tiglao, "Seaside Boom," *Far Eastern Economic Review* (July 8, 199), 14.

6. For detailed assessments of the conflicting claims to the Spratly Islands and the waters surrounding them, see Bob Catley and Makmur Keliat, *Spratlys: The Dispute in the South China Sea,* (Aldershot, U.K.: Ashgate, 1997), 24–110; and Bradford L. Thomas and Daniel J. Dzurek, "The Spratly Islands Dispute," *Geopolitics and International Boundaries* 1 (Winter 1996), 302–310.
7. Brunei has made a maritime claim that extends into the Spratly Islands area. However, its position is somewhat ambiguous and Brunei does not have troops or personnel occupying any of the Spratly Islands features as of this writing. Bradford L. Thomas and Daniel J. Dzurek, "The Spratly Islands Dispute," *Geopolitics and International Boundaries* 1 (Winter 1996), 306.
8. Translation of Beijing publication, *Renmin Ribao,* in Foreign Broadcast Information Service (FBIS), "RENMIN RIBAO Reporters Tour Spratlys," *Daily Report: China* (FBIS-CHI-95-141, July 24, 1995), 37.
9. See Mark J. Valencia, *China and the South China Sea Disputes,* Adelphi Paper 298 (London, U.K.: International Institute for Strategic Studies, 1995), 8–11; and Bob Catley and Makmur Keliat, *Spratlys: The Dispute in the South China Sea,* (Aldershot, U.K.: Ashgate, 1997), 44–65.
10. For discussions on how the Law of the Sea Convention has influenced the competing claims for the Spratly Islands, see Bradford L. Thomas and Daniel J. Dzurek, "The Spratly Islands Dispute," *Geopolitics and International Boundaries* 1 (Winter 1996), 307–310; and Lee G. Cordner, "The Spratly Islands Dispute and the Law of the Sea," *Ocean Development and International Law* 25 (January–March 1994), 61–74.
11. Alan Dupont, *The Environment and Security in Pacific Asia,* Adelphi Paper 319 (London, U.K.: International Institute for Strategic Studies, June 1998), 50–54.
12. Ibid.
13. Although this book focuses mainly on mitigating the risk of conflict over the Spratly Islands, it is worth noting that other regions of the South China Sea have been flashpoints for conflict at different times. Perhaps the most well-known is the Paracels, a small group of islands in the northern region of the South China Sea that received international attention when Chinese troops occupied them in early 1974

after forcibly ousting their South Vietnamese garrisons. The clash over the Paracels appears to offer certain similarities to the Spratly Islands disputes in terms of conflicting territorial claims, energy interests at stake, and the potential for military confrontation. For background, see Marwyn S. Samuels, *Contest for the South China Sea* (New York: Methuen, 1982), 98–103.

14. The Falklands conflict also suggests the importance of seemingly unimportant maritime locations. It is noteworthy that the key precipitating incident in escalating the longstanding diplomatic dispute between Argentina and the United Kingdom over the sovereignty of the Falklands Islands (or Malvinas from the Argentine perspective) started with the landing of scrap-metal workers from Argentina on the even more remote, British-controlled island of South Georgia in an effort to press the sovereignty question. Regarding this incident and the broader diplomatic miscalculations on both sides, see Lawrence Freedman and Virginia Gamba-Stonehouse, *Signals of War: The Falklands Conflict of 1982* (Princeton: Princeton University Press, 1991), 35–64 and 413–417; and the case study by Don Lippincott and Gregory F. Treverton, *Falklands/Malvinas* (Cambridge: John F. Kennedy School of Government, Harvard University, 1986), 14–16.

15. Scott Snyder, *The South China Sea Disputes: Prospects for Preventive Diplomacy: A Special Report of the United States Institute of Peace* (Washington, D.C.: United States Institute of Peace, August 1996), 14.

16. A key venue for this dialogue has been the "Workshops on Managing Potential Conflict in the South China Sea," which were initiated in 1990 by Indonesia and actively supported by Canada. For a first-hand account, see Hasjim Djalal and Ian Townsend-Gault, "Managing Potential Conflicts in the South China: Informal Diplomacy for Conflict Prevention," in *Herding Cats: Multiparty Mediation in a Complex World,* Chester A. Crocker *et al., eds.* (Washington, D.C.: United States Institute of Peace, 1999), 115–117.

17. Bob Catley and Makmur Keliat, *Spratlys: The Dispute in the South China Sea* (Aldershot, U.K.: Ashgate Publishing Company, 1997), 152–153.

18. John C. Baker, Kevin M. O'Connell, and Ray A. Williamson, eds., *Commercial Observation Satellites: At the Leading Edge of Global Transparency* (Santa Monica, CA: RAND Corporation and the American Society for Photogrammetry and Remote Sensing, 2001), 3–4 and 579-587.

Part

I

2

The Spratly Islands Dispute: Legal Issues and Prospects for Diplomatic Accommodation

Christopher C. Joyner

INTRODUCTION

The Spratly Islands remain a dangerous security flashpoint in Southeast Asia. Tensions over these land features present serious potential for an outright, though almost surely limited military clash, between China and other states in the region. To preclude this possibility, efforts are being undertaken to engage China in multilateral regional diplomacy, aimed at fostering discussions that might mitigate sensitive issues pertaining to the South China Sea. Through these efforts, opportunities for conflict might be short-circuited. This chapter examines such considerations.

The Spratly Islands contain some several hundred scattered small islets, sandbars, shoals, banks, atolls, cays, and reefs in the South China Sea.[1] With elevations ranging from two to six meters, the mapped features of the Spratly archipelago, including shallow territorial waters, cover an ocean area of approximately 180,000 square kilometers.[2]

The Spratlys are too small and barren to support permanent human settlement, and few have fresh water or any significant land-based resources.[3] Yet, these tiny land formations are still considered strategic, economic, and political assets for the littoral states in the South China Sea—at the least because they can serve as legal base points for states to project claims of exclusive jurisdiction over waters and resources in the South China Sea. More significantly, the Spratlys' area holds strategic importance, as these features straddle the sea lanes through which commercial vessels must sail en route to and from Southeast Asian ports.

LEGAL CONSIDERATIONS

Six states have claims to the Spratlys: China, Taiwan, and Vietnam claim the entire archipelago, while the Philippines, Malaysia, and Brunei claim a select number of the islets. All the claimant states, except for Brunei, have established a military presence in the Spratlys.[4]

National rationales to claim the Spratlys reside mainly in jurisdictional opportunities created for coastal states over offshore seabed resources as set out in the 1982 United Nations Convention on the Law of the Sea (UNCLOS).[5] This instrument codifies new rights for states having territorial sovereignty over an island or group of islands.[6] Chief among these is the exclusive right to exploit living and nonliving resources of the water column and seabed surrounding an island or archipelago. A state holding valid, legal title to sovereignty over an island is permitted by UNCLOS to establish a twelve-mile territorial sea and a 200-nautical mile exclusive economic zone (EEZ) around that island.[7] If an entire island group obtains recognized sovereign independence as an archipelagic state, it has the right to draw straight baselines between the outermost islands and will acquire exclusive rights to explore and exploit living and non-living resources within the area enclosed by that baseline as if the area were internal waters.[8] The legal rights to exploit resources offshore for non-state archipelagoes, as is the case of the Spratlys, flow from the rights to exploit continental shelves offshore groups of islands.[9]

Should all the claimant governments declare exclusive economic zones or continental shelf delimitations seaward from points fixed by those Spratly features over which they now assert sovereignty, practically the entire ocean and seabed in the South China Sea would be crisscrossed by conflicting claims of national jurisdiction. This would convert much of that ocean region, now legally comprised of high seas and international seabed area, into a semi-enclosed sea.[10]

National claims to sovereignty over the Spratlys also rest on acts of discovery and occupation. Legal considerations for the Spratlys became more salient for governments as speculation arose over the potential for petroleum deposits in local continental shelves, coupled with the emergence in 1982 of UNCLOS as the standard for demarcating offshore jurisdictional limits for resource exploitation.

China's sovereignty assertions to the Spratlys rest on historical claims of discovery and occupation. The Chinese case is amply documented, going back to references made in Chou Ch'ufei's *Ling-Wai-tai-ta* (Information on What Lies Beyond the Passes) during the Sung dynasty (twelfth century)[11] and in the records of Chinese navigators during the Qing dynasty (eighteenth century).[12] Modern international law, however, holds that merely discovering some territory is not sufficient to vest in the discoverer valid title of ownership to a territory. Rather, discovery only creates inchoate title, which must be perfected by subsequent continuous and effective acts of occupation, generally construed to mean permanent settlement.[13] Evidence of such permanent settlement is not compelling in China's claim to the Spratlys.[14] In 1992, China passed a special territorial sea and contiguous zone act to legalize its claims to the Spratlys. Article 2 of this legislation specifically identifies the Spratly (and Paracel) archipelagoes as Chinese territory.[15] To demonstrate this claim to title, since 1988 China has deployed some marines in garrisons on seven Spratly islets.[16]

The claim by Taiwan replicates that of the People's Republic of China (PRC), as its legal basis for sovereignty rests on longstanding historic ties to the islands.[17] Thus, Taiwan's claim suffers from the same deficiencies as that of the PRC, namely, that discovery of, and intermittent contact with, scattered islet formations in ocean space are insufficient cause to establish legal title to sovereignty.

Taiwan was the first government to establish a physical presence on a Spratly feature. Taiwan announced its claim to the atoll in 1947 and has occupied the largest island in the Spratlys, Itu Aba, constantly since 1956. Paradoxically, in terms of international law, such an unchallenged exercise of control over Itu Aba for more than four decades may qualify as a display of continuous and peaceful sovereignty, a condition necessary for supporting a legal claim to the island. Taiwan maintained a force of some 500 soldiers on Itu Aba from the mid-1950s through the late 1990s, later reduced to about 100 personnel.[18]

The legal grounds for Vietnam's sovereign claims to the Spratlys flow from historic activities during the Nguyen dynasty (seventeenth to nineteenth centuries). Maps and other supporting historical evidence for Vietnam's claims were compiled by the government in two white papers, *Vietnam's Sovereignty Over the Hoang Sa and Trung Sa Archipelagoes*, which were issued in 1979 and 1982. Vietnam also bases its title to the Spratlys by right of cession from a French claim to the islands made in 1933. The French, however, did not attempt to perfect its claim to title to the Spratlys by occupation.[19] Consequently, France possessed no lawful title to the Spratly group to which Vietnam could succeed. In any event, Vietnam moved in 1975 to secure its claim of possession of the Spratlys when it occupied 13 islands of the group. In September of 1989, Vietnam occupied three more islets, and has since taken at least nine additional atolls. By 2000, Vietnam had stationed 600 troops on at least 27 Spratly land formations.[20]

The Philippines justifies its Spratlys claims principally on the "discovery" of certain islands by Thomas Cloma in 1947.[21] In 1978, the Marcos government formally annexed the archipelago to the Philippines and placed it under the administration of Palawan province.[22] This Philippine claim rests on a geological assertion that the continental shelf of the so-called Kalayaan Island group is juxtaposed to the Palawan Province and extends some 300 miles westward, into the heart of the Philippines' exclusive economic zone.[23] To defend its Spratlys claims, the Philippines currently has marines stationed on eight islands. These bases are fortified with heavy artillery and are equipped with radar facilities, a weather station, and ammunition depots.[24]

More recently, Malaysia and Brunei have asserted claims to certain islands and reefs in the Spratlys, based principally on certain continental shelf provisions in the 1982 UNCLOS. These provisions describe in detail what legally constitutes a "continental shelf" for a state, and the sovereign rights a state may exercise for the purposes of exploring and exploiting the resources of its continental shelf.[25]

Malaysia has claimed sovereignty over twelve islands in the Spratly group, but those claims appear ill-founded. Serious doubt remains about the legal propriety of Malaysia's assertions, which arises from Malaysia's basing its claims to certain islands on ocean law principles associated with prolongation of a continental shelf seaward, rather than the accepted legal means of validating claim to title over territory through permanent occupation.

The clear inference from Malaysia's claims is that a state possessing a continental shelf also possesses sovereign rights over land formations arising seaward from that shelf. That inference lacks acceptance under contemporary international law. The 1982 UNCLOS neither stipulates nor invites such an interpretation. It does set out a regime for an island, which is defined as a "naturally formed area of land, surrounded by water, which is above water at high tide."[26] The Convention also gives to a state with established sovereignty over an island the right to exploit living and non-living resources in the water column and on the seabed within that island's territorial sea, contiguous zone, and exclusive economic zone. The critical legal consideration for acquisition of sovereign title over an island formation, however, is not the geological affinity of a coastal state to island formations arising from continental shelves offshore. Rather, ownership derives from occupation, demonstrated by a continuous and effective display of sovereignty through permanent settlement.[27]

Malaysia is the most recent claimant to occupy part of the Spratlys militarily. In late 1977, Malaysian troops landed on Swallow Reef. Since then, 70 soldiers have been stationed on four of the twelve islets claimed by Malaysia.[28]

Brunei has only one claim to the Spratly group, that to a naturally submerged formation known as Louisa Reef. The legal premise for substantiating Brunei's claim flows from continental shelf provisions in the 1982 UNCLOS. Louisa Reef is a submarine feature, part of the seabed, and hence legally an extension of a continental shelf. The key criterion for ownership is whether the continental shelf can be substantiated as a natural prolongation seaward from the coastal territory of Brunei. Granting that, Brunei would enjoy the exclusive right to exploit resources of the reef. Brunei remains the only claimant without a military presence in the Spratly Islands. Even so, Louisa Reef also is claimed by Malaysia, which took possession of it in 1984.[29]

Thus, the Spratlys situation remains complicated by competing claims and the possibility of military clashes. Taiwan remains in control of Itu Aba Island; the PRC has occupied seven reefs and rocks since January 1988; Vietnam now occupies at least 27 islands, reefs, and cays. The Philippines controls eight principal islands and claims some 50 other islets, reefs, and shoals. Malaysia has troops on four atolls and asserts claims to several other geological formations in the area. Brunei claims Louisa Reef. Consequently, the South China Sea has become a patchwork of conflicting national claims, most recently driven by geopolitical considerations over the speculation of potential hydrocarbon resources around the Spratlys.[30]

The contentious nature of jurisdictional disputes over the Spratlys has prompted claimant states to bolster their claims by stationing a permanent military presence in the archipelago. By 2000, over 1500 troops of five claimant governments had occupied at least 46 of the key land and marine formations in the Spratly archipelago.[31] In the process, the two principal antagonists, China and Vietnam, have each increased naval patrols and established new military outposts on previously unoccupied islets in the region.[32]

Several bilateral incidents during the 1980s and 1990s contributed to uneasy, tense feelings among various claimants. Indeed, these incidents spotlight the Spratlys situation as a contentious flash point, as seen in the 1980s when China clashed with Vietnam on a number of occasions,[33] and during the late 1990s as tensions flared between China and the Philippines over the PRC's insistent occupation of Mischief Reef, a tiny islet within the Philippines' EEZ.[34] Also disturbing is that ASEAN appears vulnerable to pressures from the PRC. The region's financial crisis strained economic relations among member states and left them with few funds to spend on military defense. Although four of the six claimant states to the Spratlys are ASEAN members (the Philippines, Malaysia, Vietnam, and Brunei), none are anxious to upset its bilateral relations with China given the latter's economic growth and potential marketplace. The point here seems obvious: The reluctance of ASEAN to stand up to recent Chinese intrusions into the South China Sea warrants serious concern. The message sent to Beijing from ASEAN's non-response is that further expansionary activities will not be seriously criticized, much less challenged by that forum.[35]

THE SOVEREIGNTY OBSTACLE

Regional efforts to deal with sovereignty disputes in the South China Sea have met with little success. China opposes multilateral talks on the Spratlys because its sovereignty over the islands is deemed non-negotiable, although joint ventures for exploiting natural resources in the area can be negotiated on a bilateral basis. China's no-multilateral negotiation strategy also is driven by strategic bargaining preferences. The PRC views its interests in bilateral terms because that strategy makes it easier to isolate the disputants and deal with them one-on-one. This erodes the ability of ASEAN to organize around an issue, while permitting China the freedom to negotiate individually with governments in that region.[36]

Sovereignty connotes legal and political dimensions. For China and Vietnam, notions of political sovereignty are sensitive concerns. Any challenge to China's claim to the Spratlys is considered to be a challenge to China's domestic sovereignty. Any concession is seen as appeasement, with adverse implications both for domestic politics and foreign relations. This point is reinforced by the realization that nationalism and sovereignty remain strong domestic political forces, cementing together the Chinese Communist Party in the post-Cold War era. In the past, the Chinese political leadership could point to foreign intervention, the Soviet threat, and irredentism to bolster its nationalist legitimacy. Today, such appeals to external anxieties by the leadership hold less political sway for challenges to Chinese claims to the Spratlys. The importance of the region is viewed more in terms of geopolitical attributes, that is, in terms of fishery resources, hydrocarbon potential, and commercial sea lanes. Thus, Sino-Vietnamese contention over the Spratlys turns less on ideology and more on access to resources, both for food and development. For China, then, control over the Spratlys can not be handed over to any adversary, especially to its principal antagonist, Vietnam.

Especially delicate for the PRC in any multilateral discussions is the issue of Taiwan. Even if both governments agreed to participate, their respective positions must be reconciled within the format of discussions. Both the People's Republic of China and the Republic of China on Taiwan claim all the Spratly formations as their sovereign territory. Both governments make identical claims to the Spratlys in the name of "China," based on similar historical evidence. Were each to enter into formal negotiations as two separate, contending parties, that would constitute de facto recognition of two Chinas, which neither government accepts. Moreover, two separate Chinas would confound claims by both governments to the Spratlys and incite squabbles over the recognition of credentials. Neither is the time ripe to put together a single, combined Chinese negotiating delegation and crystallize a common position on the Spratlys. PRC conducting combat naval maneuvers in the Taiwanese Strait in 1996 seriously aggravated political relations and dimmed any trust between Beijing and Taipei. For its part, the PRC endorses diplomatic negotiation in principle, but prefers limited bilateral approaches in fact.[37]

CONFIDENCE-BUILDING MEASURES

For international cooperation over South China Sea issues, negotiations must be forthcoming among the Spratly claimants. The missing but clearly necessary ingredients for these negotiations are sustained confidence and transparency between China and the rest of Southeast Asia. An important first step for these governments is to engage in confidence-building measures (CBMs). Through CBMs, functional cooperation and direct communication might be fostered among the claimants to preclude disagreements or misperceptions from escalating into military confrontation. Measures for confidence-building can produce a better climate for negotiations.

Building confidence depends on nurturing mutual trust and understanding. For genuine confidence to be secured, governments must understand the motives and rationales behind the policies of other states in the region, which can only evolve through increased transparency of national policies and intentions. Governments must also understand that competing self interests and violations of past agreements carry high political costs, as they dissipate mutual trust in which CBMs are grounded. Transparency thus becomes fundamental to confidence-building. In the case of the South China Sea, the following are a number of measures to enhance transparency and thus build confidence among the concerned governments:

- Spratly claimant governments might consider giving official and informal assurances to restrain the use of military force in the region.
- Government officials might strive to recognize and respect national sensitivities arising from military deployments in the region. While some Spratly claimants might prefer not to recognize the legitimacy of other governments' claims in the region, they must respect the sensitivities arising from those claims.
- Claimant governments must cease further occupation and annexation of territory in the Spratlys. Seizing and occupying more islets does little to promote a government's strategic position in the region, and new occupations reinforce suspicion and distrust over that government's disingenuousness toward future diplomatic negotiations. It seems prudent that claimant governments should accept the status quo as the starting ground for negotiation.
- Governments claiming the South China Sea territories might reign in efforts to expand military activities in the region. Military activities connote shows of force, not genuine sincerity to resolve contentious legal issues peacefully.

- A common set of operating procedures for the Navies and Air Forces of concerned governments in the disputed region would be helpful. Such a "standardized manual of operations" would lessen tensions by reducing the likelihood for accidents and minimizing situations that could spark military conflict in the region.
- Efforts might be made to devise means and mechanisms that improve contacts and communication between mainland governments and their local military commanders on islands occupied in the Spratlys region. Again, establishing state-to-state hotlines of communication could reduce the chances of misunderstandings and misperceptions of other governments' policy intentions.[38] The establishment of hotlines between naval chiefs might also be useful.

The process of confidence-building amongst the Spratly governments has already begun. A regional dialogue on disputes, hosted informally by Indonesia through a series of workshops, has been convened annually since 1990. Although Indonesia is a South China Sea littoral state, it is usually viewed as a neutral party in the region, as it makes no claim to the Spratly archipelago.[39] Its motive for undertaking this initiative appears straightforward: If regional tensions can be reduced and greater peace and order brought to the South China Sea, then Indonesia will share in the resultant growth in economic and commercial opportunities.

The Indonesian initiative, also labeled the South China Sea Informal Working Group, strives to encourage confidence among South China Sea states through "track two diplomacy" so as to ease tensions arising from sovereignty and jurisdictional disputes over the Spratly (and Paracel) Islands. The strategy here is that persons from various states party to disputes in the region meet informally and discuss facets and issues of the matter, in order to create an atmosphere of open free discussion, without having the restrictions implicit in having to maintain official government positions.[40] Eleven workshops have been held through December 2000, guided increasingly with the pragmatic realization that the activities discussed in these meetings must be done on a truly regional basis.

The Indonesian workshops represent the most serious effort so far for promoting peace and cooperation in the South China Sea. These meetings serve as informal, private fora for confidence-building among nationals from states involved in Spratly Islands jurisdictional disputes. They have been purposefully designed to bring together representatives from concerned states in the region to discuss a variety of non-polemical issues. Special working groups and collaborative projects have been organized, with a view to promoting regional cooperation on the marine environmental issues, including *inter alia*, biodiversity, pollution control, hydrographic information and data exchange, marine research, habitat protection, and cooperation on search and rescue operations, and in dealing with illegal acts at sea, including piracy.

The PRC's attitude toward this deliberation process is critical.[41] China has supported the workshop process, although Beijing apparently feels that the process is going too far, too fast. For China, protracted patience is diplomatic virtue for formulating arrangements affecting the Spratlys. The Chinese have supported efforts to promote cooperation on select issues (such as marine scientific research, marine environmental protection, and safety and sea communications), though it is less clear whether China will implement any of the agreed upon collaborative projects.[42] Similarly, the PRC acknowledges the need to develop confidence building among states in the region, but seems to regard the workshop process as that end in itself. China is reluctant to discuss confidence-building measures that relate to national sovereignty or territorial claims in the region, which it feels lie beyond the competence of the workshop. Importantly, China has intimated a willingness in principle to set aside territorial claims in favor of joint development. Left unclear, however, is the location and meaning of what "zone" China might be willing to explore or develop jointly. What has been made clear, however, is that "joint development" means to China precisely that: Bilateral development undertaken jointly with another claimant in an area of the South China Sea claimed by the other state.[43] Resource development is a bilateral process, not a multilateral enterprise in which all the Spratly claimants are involved. Even so, on various activities affecting the marine environment of the South China Sea, Beijing appears to be moving more toward acceptance of a regional approach.

Given this series of Indonesian workshop sessions, a strategy of confidence building is in progress and has produced results—at least in terms of the number of ancillary discussions being convened among the participants. But the process is slow, ponderous, and piecemeal. While achieving few visible results, the workshop sessions and their spinoff committee discussions do provide opportunities for participants to share their views, thus compelling persons from claimant governments to recognize differences of opinion. The point here thus far is that the workshops reveal a process toward regional cooperation, not a quick fix for demilitarization or the relaxation of sovereignty claims in the South China Sea.

These concerns aside, current relations among the Spratly claimants seem relatively good. Governments in Southeast Asia continue to move toward patterns of accommodation and normalization. China, at least rhetorically, indicates a willingness to cooperate with other states on joint development in the Spratlys.[44] Recent events suggest that while no negotiation means no compromise on Chinese sovereignty over the archipelago, resources might be exploited by other claimants if the PRC does nothing to protect those resources. Diplomatic pressures by a united ASEAN could work to dissuade China from taking a military response, largely because such hostilities would unravel increasingly cooperative commercial relations in the region.

The ultimate objective for the Spratly Islands should be strategic stability, national security, and reduced force levels for all states in the region. There is a need for more non-military mechanisms to keep the peace amongst interested governments. Multilateral fora, such as the Asia-Pacific Economic Cooperation Conference, are important institutions for promoting transparency and mutual exchange of information on regional activities. Through such agreed upon rules, tension and conflict among the Spratly claimant states might be better contained.[45]

CONCLUSIONS

The core of the Spratly dispute turns on questions of territorial sovereignty, not law-of-the-sea issues. The complexities of overlapping maritime claims and the dispute's long history make determination of national sovereignty in the Spratlys extremely difficult. Obviously, if the sensitive issue of sovereignty can be redressed, or at least legally shelved as a divisive nationalistic factor, then the maritime jurisdictional principles codified in the 1982 UNCLOS can be more aptly applied to the Spratlys group. Such application could cede undersea resource rights to portions of the South China Sea to recognized legal owners. Yet, no claimant government today can establish sufficiently substantial legal grounds to validate its claim in the eyes of the international community.

ASEAN states remain particularly concerned because China historically tends to resort to military force to settle disputes within what it regards as its national sphere of influence. The PRC's build-up of naval forces is seen as expanding this sphere in the South China Sea through enhancing force projection, and its move to a blue-water navy appears as spawning a naval arms race among Asian states. China's vast size, growing population, rapidly expanding economy, and improved air and naval force projection capabilities make it increasingly formidable as a regional power, and its policy toward the Spratlys reflects that attitude.

To conclude, these are the facts: In 2001, no disputant in the region possessed the ability to fight a sustained war over these islets. Further, few of the land formations are sufficiently large to build airstrips. At the same time, the resource potential in the region remains unknown, though speculation abounds. The essential point is that the time for building trust, transparency, and cooperation is now. Without preexisting institutional arrangements, should some exploration venture make a major oil strike, or if China were to acquire aircraft carrier capabilities in the South China Sea, beggar-thy-neighbor, self-help policies might well prevail over regional efforts at peaceful accommodation.

There is still need for an acceptable diplomatic solution for the Spratlys. Conflicting claims over the Spratlys are driven by domestic politics. Irrespective of the talents and personalities of negotiators around a table, if government officials refuse to exercise the political will to produce a solution, such an outcome cannot happen. If leaders are not willing to compromise on sovereignty issues to reach a long-term agreement, then no agreement will be made. Governments must genuinely desire to manage the dispute—to negotiate an agreement that brings acceptable benefits to each party, though not at the expense of any vital interests to another.

Conflict over the Spratlys can be managed only if policy makers demonstrate the political will and genuine determination to do so. Most states in Southeast Asia confront common problems, among them political succession, economic development, and rising expectations from their people. In this regard, the Spratlys situation could be converted into an issue of regional unity, rather than divisiveness, through a resort to CBMs and diplomatic accommodation. That ambition has been successfully obtained for other contentious multinational sovereignty disputes elsewhere,[46] and such political and legal strategy for the South China Sea seems no less plausible. The critical challenge is for the PRC and other governments in the region to make it happen.

NOTES

1. "ICE Cases: Spratly Islands Dispute," Case No. 21 (May 1997), located at http://gurukul.ucc.american.edu/ted/ice/Spratly.htm; and Dieter Heinzig, *Disputed Islands in the South China Sea* (Hamburg: Institute of Asian Affairs, 1976).

2. J. K. T. Chao, "South China Sea: Boundary Problems Relating to the Nansha and Hsisha Islands," in R. D. Hill, Norman G. Owen, and E. V. Roberts, eds., *Fishing in Troubled Waters: Proceedings of an Academic Conference on Territorial Claims in the South China Sea* (Hong Kong: Center of Asian Studies, 1991), 77. The total number of the Spratly Islands, reefs, shoals, and cays is variously reported. See J. R. V. Prescott, *The Maritime Political Boundaries of the World* (London: Methuen, 1985), 218–219; Mark J. Valencia, Jon M. Van Dyke, and Noel A. Ludwig, *Sharing the Resources of the South China Sea* (Hague: Martinus Nijhoff Publishers, 1997), 225–235 (Appendix 1).

3. See generally David Hancox and Victor Prescott, "A Geographical Description of the Spratly Islands and an Account of Hydrographic Surveys Amongst Those Islands," *Maritime Briefing* 1 (1995) (International Boundaries Research Unit, special issue); and Tao Cheng, "The Dispute Over the South China Sea Islands," *Texas International Law Journal* 10 (1975), 267.

4. See Cheng-yi Lin, "Taiwan's South China Sea Policy," *Asian Survey* 37 (1997), 324 (Table 1); Mark J. Valencia, *China and the South China Sea Disputes*, Adelphi Paper 298 (Oxford, U.K.: International Institute for Strategic Studies, 1995), 8–25; Mark J. Valencia, *South-East Asian Seas, Oil under Troubled Waters: Hydrocarbon Potential, Jurisdictional Issues and International Relations* (Singapore: Oxford University Press, 1985), 87–89. For historical assessments of claims to the Spratlys, see Steven Kuan-tsyh Yu, "Who Owns the Paracels and Spratlys? —An Analysis of the Nature and Conflicting Territorial Claims," *Chinese Yearbook of International Law and Affairs* 9 (1989–1990), 1–27 (substantiating the Chinese claims).
5. United Nations Convention on the Law of the Sea, opened for signature December 10, 1982, entered into force November 16, 1994. UN Doc. A/CONF.62/122 (1982), reprinted in United Nations, *The Law of the Sea: Official Text of the United Nations Convention on the Law of the Sea with Annexes and Index*, UN Sales No. E.83.V.5 (1983) [hereinafter 1982 UNCLOS].
6. Ibid., Articles 121 and 46–54.
7. Article 121 provides in relevant part that "...the territorial sea, the contiguous zone, the exclusive economic zone and the continental shelf of an island are determined in accordance with the provisions of this Convention applicable to other land territory." Ibid., Article 121 (2).
8. Ibid., Articles 46–54.
9. Ibid., Articles 77 and 81.
10. See Mark J. Valencia, Jon M. Van Dyke, and Noel A. Ludwig, *Sharing Resources of the South China Sea* (Hague: Martinus Nijhoff Publishers, 1997) 254 (Plate 1). For background assessments, see Choon-Ho Park, "The South China Sea Disputes: Who Owns the Islands and the Natural Resources?" *Ocean Development and International Law* 1 (1973), 37; and Geraldo M. C. Valero, "Spratly archipelago dispute: Is the question of sovereignty still relevant?," *Marine Policy* 18 (1994), 314–344.
11. Chou Ch'u-fei, *Ling-Wai-tai-ta* (Information on What Lies Beyond the Passes, 1178), cited in Marwyn S. Samuels, *Contest for the South China Sea* (New York: Methuen, 1982), 15–16; and Shao Hsun-Cheng, "Chinese Islands in the South China Sea," *People's China* 13 (1956), 26.
12. See Marwyn S. Samuels, *Contest for the South China Sea*, Appendix C. Also see generally Z. Han, J. Lin, and F. Wu, eds., *Collection of Historical Materials on China's Islands in the South China Sea* (Beijing:

Oriental Press, 1988); and Jianming Shen, "Territorial Aspects of the South China Sea Disputes," in Myron H. Norquist and John Norton Moore, eds., *Security Flashpoints: Oil, Islands, Sea Access and Military Confrontation* (Hague: Martinus Nijhoff, 1998), 139–218.

13. *Legal Status of Eastern Greenland Case (Denmark v. Norway), Permanent Court of International Justice*, Series A/B, No. 53 (1933), 3 Hudson, *World Court Reports*, 148.
14. See generally Harry L. Roque, Jr., "China's Claim to the Spratly Islands under International Law," *Journal of Energy & Natural Resources Law* 15 (1997), 189–211; and Michael Bennett, "The People's Republic of China and the Use of International Law in the Spratly Islands Dispute," *Stanford Journal of International Law* 28 (1997), 425–450.
15. The Law of the People's Republic of China on the Territorial Sea and Contiguous Zone became effective on February 25, 1992. See *People's Daily* (Beijing, February 26, 1992), 4, reprinted in United Nations, *Law of the Sea Bulletin No. 21* (August 1992), 24–27. For discussion of this law's legal ramifications, see Liyu Wang and Peter H. Pearse, "The New Legal Regime for China's Territorial Sea," *Ocean Development and International Law* 25 (1994), 431–442; and Nien-Tsu Alfred Hu, "The Two Chinese Territorial Sea Laws: Their Implications and Comparisons," *Ocean and Coastal Management* 20 (1995), 89–96.
16. See Table 1.1.
17. See Peter Kien-hong Yu, "Reasons for Not Negotiating on the Spratlys: A Chinese View from Taiwan," in R. D. Hill *et al.*, *Fishing in Troubled Waters*, 139–149.
18. See Table 1.1, and Cheng-yi Lin, "Taiwan's South China Sea Policy," 324. Cf. *Central Daily News* (Taiwan, December 2, 1992), 4.
19. Nor did the French act by returning to the islands after Japan's departure following World War II, or by acting to assert sovereignty after Japan formally relinquished all title and future claims to the islands at the San Francisco Conference of 1951. See Hungdah Chiu and Choon-Ho Park, "Legal Status of the Paracel and Spratly Islands," *Ocean Development and International Law* 3 (1975), 18; and Lee G. Cordner, "The Spratly Islands Dispute and the Law of the Sea," *Ocean Development and International Law* 25 (1994), 65.
20. See Table 1.1.
21. Thomas Cloma, an enterprising Filipino businessman and owner of a fishing fleet and private maritime training institute, aspired to open a cannery and develop guano deposits in the Spratlys. It was principally

for economic reasons, therefore, that he "discovered" and claimed the Kalayaan Islands and established several colonies on them by 1950. Marwyn S. Samuels, *Contest for the South China Sea* (New York: Methuen, 1982), 81–86.

22. See Presidential Decree No. 1596, "Declaring Certain Areas Part of the Philippine Territory and Providing for Their Government and Administration, June 11, 1978."
23. See Gil S. Fernandez, "The Philippines' South China Sea Island Claims," in Aileen San Pablo-Baviera, ed., *The South China Sea Disputes: Philippine Perspectives* (Manila: Philippine-China Development Resource Center, 1992), 20.
24. See Table 1.1.
25. 1982 LOS Convention, Articles 76 and 77.
26. Ibid., Article 121 (1).
27. "Rocks" in the ocean that cannot sustain life are not islands and cannot generate offshore jurisdictional zones. LOS Convention, Article 121(3). See Jon I. Charney, "Rocks that Cannot Sustain Human Habitation," *American Journal of International Law* 93 (1999), 863–877.
28. See Table 1.1.
29. See generally Khadijah Muhamed and Tunku Shamsul Bahrin, "Scramble for the South China Sea: The Malaysian Perspective," in R. D. Hill *et al.*, *Fishing in Troubled Waters*, 237–250.
30. See Bob Catley and Makmur Keliat, *Spratlys: The Disputes in the South China Sea* (Aldershot: Ashgate Publishing Company, 1997), 44–65. For the evolution of these claims, see Tao Cheng, "The Dispute Over the South China Sea Islands," 265; see Hungdah Chiu and Choon-Ho Park, "Legal Status of the Paracel and Spratly Islands," *Ocean Development and International Law* 3 (1975), 1. Importantly, there is actually little evidence to support the speculation of extensive potential hydrocarbon reserves in the region. See U.S. Energy Information Service, South China Sea Region (Washington, D.C.: U.S. Energy Information Service, 1998), located at www.eia.doe.gov/emeu/cabs/schina.html, 3–4 (as of August 2000).
31. Many of these are not military bases as such, but rather raised manmade platforms constructed on and secured to barren projections of natural coral outcroppings. Such structures are highly vulnerable to sea and weather conditions, and must be continually resupplied because there is a dearth of fresh water available. For photographs of these structures, see Mark J. Valencia, Jon M. Van Dyke, and Noel A. Ludwig, *Sharing Resources of the South China Sea*, (Hague: Martinus

Nijhoff Publishers, 1997) 258 (Plate 5); *Asian Defense Journal* 15 (1992), 22; and Christopher C. Joyner, "The Spratly Island Dispute in the South China Sea," *Integrated Coastal Zone Management* (Spring 2000), 85–90.

32. See Table 1.1.

33. The first such clash happened on February 8, 1987, when Chinese and Vietnamese warships in the area exchanged fire. On March 14, a more serious confrontation occurred off Union Reef, as each navy lost a vessel and 120 Vietnamese sailors drowned. Peter Forrest and Eric Morris, "Maritime Constabulary and Exclusive Economic Zones in the South China Sea: Some Strategic Considerations," in R. D. Hill *et al.*, *Fishing in Troubled Waters*, 302–319, 311. Another violent clash between China and Vietnam occurred in March 1988 on Fiery Cross Reef (Chigua atoll). See Chang Pao-Min, "A New Scramble for the South Sea Islands," *Contemporary Southeast Asia* 12 (June 1990), 20–39. After a small contingent of Vietnamese fired on Chinese military and construction personnel working there, the Chinese dispatched warships to the area and sunk three Vietnamese vessels, with the loss of 74 lives. See *Straits Times* (March 26, 1988), 3. China emerged as the clear victor from this episode. Not only did the Fiery Cross Reef clash demonstrate the superiority of Chinese naval power over that of Vietnam, it also reaffirmed the PRC's determination to assert sovereignty over the Spratlys. See Justus M. van der Kroef, "Territorial Claims in the South China Sea: A Strategic Irrelevance?" in R. D. Hill *et al.*, *Fishing in Troubled Waters*, 21–35; and Shee Poon Kim, "The March 1988 Skirmish over the Spratly Islands and Its Implications for Sino-Vietnamese Relations," in Ibid., 177–191.

34. Greater concern about Chinese intentions for the Spratlys arose during the 1990s. In 1992, China installed sovereignty markers on various shoals and islets in the Spratlys, but in response to a strong Declaration on South China Sea by ASEAN, Beijing tempered those activities over the next two years. China's expansionist policies resumed in February 1995 when its troops occupied Mischief Reef, an aptly named shoal in the Spratlys located well inside the 200-nautical mile EEZ claimed by the Philippines. See Daniel J. Dzurek, "China Occupies Mischief Reef in Latest Spratly Gambit," *Boundary and Security Bulletin* 3 (April 1995), 65–71. The PRC constructed three "fisherman's structures" on the half-submerged atoll. The Mischief Reef incident so alarmed the ASEAN states that they took a concerted stand to protest China's occupation of territory within the Philippines' maritime zone. In response to this ASEAN action, China and the Philippines agreed in 1995 to a Code of Conduct to avoid new provocations and potentially

destabilizing actions. That Code sought to reduce opportunities for military conflict over the Spratlys by encouraging force reductions in the region. See "Manila, Beijing agree on Spratlys Code of Conduct," *Reuters* (August 10, 1995). China also pledged to abide by the 1992 ASEAN statement, which called for mutual restraint in South China Sea activities. It was the crisis in 1995 over Mischief Reef, however, that provoked acute suspicion over China's long-term aims and geostrategic objectives in the Spratlys. See Scott Snyder, *The South China Sea Dispute: Prospects for Preventive Diplomacy: A Special Report of the United States Institute of Peace* (Washington D.C.: United States Institute of Peace, August 1996), 15. It also revived concerns that the South China Sea could become a tinderbox in the region. The Philippines reacted by diplomatically protesting China's "invasion of their EEZ" and sending a symbolic force of ten aircraft and three patrol boats to the island of Palawan. See Abby Tam, "Manila Tries Diplomacy In Confronting China," *Christian Scientist Monitor* (February 22, 1995). In October 1998, construction resumed on Mischief Reef, and continued through early 1999, straining relations with the Philippines. See Annie Ruth and Raffy Jimenez, "China Fortifies Hold On Spratlys," *Manila Times*, January 21, 1999. China's decision to proceed with this project demonstrates the extent to which the balance of power has tipped in China's favor since the East Asian financial crisis in 1998, which sapped the strength, attention, and unity of ASEAN members. See Rigoberto Tiglao and Andrew Sherry, "Politics & Policy," *Far Eastern Economic Review* (December 24, 1998), 18–20. Moreover, the PRC's intrusive return to Mischief Reef revived regional anxieties in the region about China. By sending warships unannounced to the Spratlys and undertaking construction operations on Mischief Reef, the PRC openly breached the 1995 Code of Conduct signed with the Philippines. Completing construction of hardened military communication facilities that are guarded by Chinese naval vessels is hardly conducive to peaceful relations between states who claim the same specks of land in the sea. Such developments strain prospects for maintaining stable regional security. See Johnna R. Villaviay, "China Eyes Another Island in the Spratlys," *Manila Times* (January 22, 1999). Similarly, serious questions are raised about the sincerity of Chinese diplomacy in light of the Mischief Reef II episode, especially given the commitment made by Jiang Zemin in November 1998 at the 6th Leaders Summit of the Asia-Pacific Economic Cooperation (AEPC) forum in Kuala Lumpur to employ peaceful means to resolve the countries' dispute over the Spratlys.

35. Ralph Cossa, "Inside View: Mischief Reef Flashpoint," *Defense News* (January 11, 1999), 13. For treatment of China's historical attitude toward ASEAN, see Bob Catley and Makmur Keliat, *Spratlys: Dispute in the South China Sea* (Aldershot: Ashgate Publishing Company, 1997), 92–110.
36. In the first Mischief Reef episode, ASEAN did put pressure on China and advocate the adoption of "codes of conduct." In the end, though, China was able to negotiate bilaterally with the Philippines to secure its objectives, and paid only lip service to the "rules of conduct."
37. The financial crisis throughout Asia in 1998 gave Taiwan the opportunity to break out of its isolation by greatly expanding commercial links throughout Southeast Asia via an investment strategy known as the "Go South" policy. By purchasing corporate and banking assets at sale prices, Taiwanese businesses are integrating their investments Asia-wide, and in the process are enhancing Taiwan's economic influence and political leverage against Beijing. See Keith B. Richburg, "Exploiting Asia's Crisis: Taiwan Buys Up Bargains And Widens Its Influence," *Washington Post* (January 2, 1998), A23. For an assessment of how the Spratly Islands dispute might be used to further normalization of relations between the PRC and Taiwan, see Christopher C. Joyner, "The Spratly Islands Dispute: What Role for Normalizing Relations between China and Taiwan?" *New England Law Review* 32 (1998), 819–852.
38. Compare the views of B. A. Hamzah, "The Spratlys: What Can Be Done to Enhance Confidence?," in R. D. Hill *et al.*, *Fishing in Troubled Waters*, 320–356.
39. Within the past few years, the PRC has made insinuations about claiming sea areas in Indonesia's exclusive economic zone around Natuna, which has pulled Indonesia directly into controversy with the PRC. This development, as seen by some observers, especially in Malaysia, may have compromised Indonesia's alleged neutrality in the region.
40. Ian Townsend-Gault, "Brokering Cooperation in the South China Sea," in Lorne K. Kriwoken, Marcus Haward, David VanderZwaag, and Bruce Davis, eds., *Oceans Law and Policy in the Post-UNCED Era: Australian and Canadian Perspectives* (Hague: Kluwer Law International 1996), 313, 316–17. For background treatment of the Indonesian initiative, see Bob Catley and Makmure Keliat, *Spratlys: Dispute in the South China Sea* (Aldershot: Ashgate Publishing Company, 1997), 139–171. Success of the Indonesian workshops depends on the participants' personalities, and the central role is being played by Indonesia's Ambassador Hasjim Djalal. He enjoys

considerable respect not only for being co-chair of the meetings, but also for his recognized stature as a diplomat of long standing, an expert on law of the sea, and the recently elected President of the International Seabed Authority. That the participants trust in and respect Ambassador Djalal contributes to his role in confidence-building.

41. Without China's intimate involvement and willingness to cooperate, solutions for the dispute over the Spratly Islands cannot be even be addressed. See generally Omar Saleem, "The Spratly Islands Dispute; China Defines the Millennium," *American University International Law Review* 15 (2000), 527–582.

42. Ibid., 128. Attitudes such as this have prompted some analysts to conclude that the Indonesian workshop mediation thus far has been a "failure." See Bob Catley and Makmur Keliat, *Spratlys: Dispute in the South China Sea* (Aldershot: Ashgate, 1997), 165–168.

43. Ibid., 128. In its bilateral overtures, China at times has demanded recognition of its sovereignty over the South China Sea in return for accepting a model of joint development. That position, of course, is unconvincing for other claimant states. See "Cautious progress on disputes affecting China seas exploration," *Offshore* (October 1998).

44. In 1993, Premier Li Peng asserted that: "On the issue of Spratly Islands whose sovereignty belong to China, our country puts forward the proposal of 'shelving disputes in favor of joint development,' and is willing to work toward the long-term stability, mutual benefit and co-operation in the South China Sea region." See "Annual Work of the Chinese Government by Premier Li Peng, to the First Plenary Session of the Eighth National People's Congress" (March 13, 1993), reprinted in *People's Daily* (April 2, 1993), 3.

45. See Peter Polomka, "Strategic Stability and the South China Sea: Beyond Geopolitics," in Hill *et al.*, *Fishing in Troubled Waters*, 36–47. For a variety of other proposals, see Mark J. Valencia, Jon M. Van Dyke, and Noel A. Ludwig, *Sharing Resources of the South China Sea* (Hague: Martinus Nijhoff Publishers, 1997), 199–225.

46. For example, special legal arrangements for the Antarctic, the Timor Gap (between Australia and Indonesia), and Spitsbergen Island had to overcome sovereignty concerns. See Christopher C. Joyner, "The Spratly Islands Dispute: Rethinking the Interplay of Law, Diplomacy, and Geopolitics in the South China Sea," *International Journal of Marine and Coastal Law* 13 (1998), 218–224.

3

Energy Issues in the South China Sea Region

Erik Kreil

INTRODUCTION

The South China Sea encompasses a portion of the Pacific Ocean stretching roughly from Singapore and the Strait of Malacca in the southwest, to the Strait of Taiwan (between Taiwan and China) in the northeast. This area includes hundreds of small islands, rocks, and reefs, with the majority located in the Paracel and Spratly Island chains. Many of these islands are partially submerged islets, rocks, and reefs that are little more than shipping hazards not suitable for habitation; the total land area of the Spratly Islands is less than three square miles. The islands are important, however, for strategic and political reasons, because ownership claims to them are used to bolster claims to the surrounding sea and its resources.

The South China Sea is rich in natural resources such as oil and natural gas. These resources have garnered attention throughout the Asia-Pacific region. Until recently, East Asia's economic growth rates had been among the highest in the world, and despite the region's 1997–1998 financial crisis, growth prospects in the long-term remain among the best in the world. This economic growth will be accompanied by an increasing demand for energy. Over the next 20 years, oil consumption in developing East Asian countries (excluding India) is expected to rise by almost three percent annually on average, with almost half of this increase coming from China. If this growth rate is maintained, oil demand for these nations will increase from about 12 million barrels per day in 2000 to more than 20 million barrels per day by 2020—a two-thirds increase over the current consumption levels.

Almost all of this additional Asian oil demand, as well as Japan's oil needs, will need to be imported from the Middle East and Africa and pass through the strategic Strait of Malacca into the South China Sea. Countries in the Asia-Pacific region depend on seaborne trade to fuel their economic growth, and this dependence has led to the sea's transformation into one of the world's busiest shipping lanes. Over half of the world's merchant fleet (by tonnage) sails through the South China Sea every year. The economic potential and geopolitical importance of the South China Sea region has resulted in jockeying among the surrounding nations to claim this sea and its resources for themselves.

SOUTH CHINA SEA TERRITORIAL ISSUES

Competing territorial claims over the South China Sea and its resources are numerous, with the most contentious revolving around the Spratly Islands and Paracel Islands. Ownership of virtually all of the South China Sea is contested. The disputed areas often involve oil and gas resources:

- Indonesia's ownership of the gas-rich fields offshore of the Natuna Islands was undisputed until China released an official map with unclear maritime boundaries indicating that Chinese-claimed waters in the South China Sea may extend into the waters around the Natuna Islands. Indonesia responded by choosing the Natuna Islands region as the site of its largest military exercises to date in 1996.
- Many of Malaysia's natural gas fields located offshore Sarawak also fall under the Chinese claim.
- The Philippines' Malampaya and Camago natural gas and condensate fields are in Chinese-claimed waters. Shell Philippines Exploration (SPEX) is developing the Malampaya fields at an estimated cost of $4.5 billion U.S., the largest single investment in the country's history. In May 2000, SPEX announced that significant amounts of oil had also been discovered at Malampaya.
- Maritime boundaries in the gas-rich Gulf of Thailand portion of the South China Sea have not been clearly defined. Several companies have signed exploration agreements but have been unable to drill in a disputed zone between Cambodia and Thailand. Overlapping claims between Thailand and Vietnam were settled on August 8, 1997, and cooperative agreements for exploration and development were signed for the Malaysia-Thai and Malaysia-Vietnam Joint Development Areas (the latter effective June 4, 1993).
- Vietnam and China have overlapping claims to undeveloped blocks off the Vietnamese coast. A block referred to by the Chinese as Wan' Bei-21 (WAB-21) west of the Spratly Islands is claimed by the

Vietnamese in their blocks 133, 134, and 135. The inability to resolve these disputes has prevented Conoco and PetroVietnam from undertaking exploration in these blocks that had been planned under a tentative pact. In addition, Vietnam's Dai Hung (Big Bear) oil field is at the boundary of waters claimed by the Chinese.

- In April 2000, China announced that a potential new source of natural gas—natural gas hydrates (frozen methane)—had been discovered for the first time in Chinese territory on the seabed around the Paracel Islands. Vietnam still recognizes the Paracel Islands as part of Vietnam.

In addition, Taiwan's claims in the South China Sea are similar to those of China. Most of the competing claims in the South China Sea are historical, but they are also based upon internationally accepted principles extending territorial claims offshore onto a country's continental shelf, as well as the 1982 UNCLOS.

UNITED NATIONS LAW OF THE SEA

The 1982 convention created a number of guidelines[1] concerning the status of islands, the continental shelf, enclosed seas, and territorial limits. Among the most relevant to the division of resources in the South China Sea are the following:

- Article 3, which establishes that "every state has the right to establish the breadth of its territorial sea up to a limit not exceeding 12 nautical miles"
- Articles 55–75 define the concept of an Exclusive Economic Zone (EEZ), which is an area up to 200 nautical miles beyond and adjacent to the territorial sea. The EEZ gives coastal states "sovereign rights for the purpose of exploring and exploiting, conserving and managing the natural resources, whether living or non-living, of the waters superjacent to" (above) "the seabed and of the seabed and its subsoil. . . ."
- Article 76 defines the continental shelf of a nation, which "comprises the seabed and subsoil of the submarine areas that extend beyond its territorial sea throughout the natural prolongation of its land territory to the outer edge of the continental margin, or to a distance of 200 nautical miles. . . ." This is important because Article 77 allows every nation to exercise "over the continental shelf sovereign rights for the purpose of exploring it and exploiting its natural resources."
- Article 121, which states that rocks that cannot sustain human habitation or economic life of their own shall have no exclusive economic zone or continental shelf

REGIONAL CONFLICT AND RESOLUTION

The establishment of the EEZ created the potential for overlapping claims in semi-enclosed seas such as the South China Sea. These claims could be extended by any nation that could establish a settlement on the islands within the region. South China Sea claimants have clashed as they have tried to establish outposts on the islands (mostly military) in order to conform with Article 121 in pressing their claims. UNCLOS states that countries with overlapping claims must resolve their claims by good faith negotiation. The use of the Joint Development Area principle, followed in the Gulf of Thailand, is one model that has been successfully used by South China Sea claimants.

All of the Spratly Island claimants have occupied some of the Spratly Islands, and/or stationed troops and built fortified structures on the reefs. Brunei, which does not claim any of the Spratly Islands, has not occupied any of them, but has declared an EEZ that includes Louisa Reef.

Military skirmishes have occurred numerous times over the past two decades, involving all five claimants to the Spratly Islands. In addition to military clashes, there also have been numerous disputes over drilling and exploration for oil and natural gas in the South China Sea (see Table 3.1).

Table 3.1

Disputes over Drilling and Exploration for Oil and Natural Gas in the South China Sea

Date	***Countries***	***Disputes***
1992	China, Vietnam	In May, China signed a contract with U.S. firm Crestone to explore for oil near the Spratly Islands in an area that Vietnam says is located on its continental shelf, over 600 miles south of China's Hainan Island. In September, Vietnam accused China of drilling for oil in Vietnamese waters in the Gulf of Tonkin.
1993	China, Vietnam	In May, Vietnam accused a Chinese seismic survey ship of interfering with British Petroleum's exploration work in Vietnamese waters. The Chinese ship left Vietnamese block 06 following the appearance of 2 Vietnamese naval ships.
1993	China, Vietnam	In December, Vietnam demanded that Crestone cancel off-shore oil development in nearby waters.
1994	China, Vietnam	Crestone joined with a Chinese partner to explore China's Wan' Bei-21 (WAB-21 block). Vietnam protested that the exploration was in Vietnamese waters in their blocks 133, 134, and 135. China offered to split Wan' Bei production with Vietnam, as long as China retained all sovereignty.

1994	China, Vietnam	In August, Vietnamese gunboats forced a Chinese exploration ship to leave an oil field in a region claimed by the Vietnamese.
1996	China, Vietnam	In April, Vietnam leased exploration blocks to U.S. firm Conoco, and ruled out cooperation with U.S. oil firms that signed Chinese exploration contracts in disputed waters. Vietnamese blocks 133 and 134 cover half the zone leased to Crestone by China. China protested and reaffirmed a national law claiming the South China Sea as its own in May.
1997	China, Vietnam	In March, Vietnam issued a protest after the Chinese Kantan-3 oil rig drilled near the Spratly Islands in March. The drilling occurred offshore of Da Nang, in an area Vietnam calls block 113. This block is located 64 nautical miles off the Chan May cape in Vietnam, and 71 nautical miles off China's Hainan Island. The diplomatic protests were followed by the departure of the Chinese rig.
1997	China, Vietnam	In December, the Vietnamese protested after the Exploration Ship No. 8 and two supply ships entered the Wan' Bei exploration block. All three vessels were escorted away by the Vietnamese navy.
1998	China, Vietnam	In September, the Vietnamese protested after a Chinese report stated that Crestone and China were continuing their survey of the Spratly Islands and the Tu Chinh region (Wan' Bei in Chinese).

OIL

The focus of most attention regarding the South China Sea's resources has been on hydrocarbons in general, and on oil in particular. Oil deposits have been found in most of the littoral (adjacent) countries of the South China Sea. The South China Sea region has proven oil reserves estimated at about 7.7 billion barrels (Table 3.2), and oil production in the region is currently over 1.8 million barrels per day. Malaysian production accounts for almost one-half of the region's total. Total South China Sea production has increased gradually over the past few years, primarily as additional production from China, Malaysia, and Vietnam came online. Compared with other offshore oil and gas regions, the South China Sea is roughly comparable to the Caspian Sea Region and Gulf of Mexico in terms of current oil production (see Table 3.3 for a comparison with other regions).

Table 3.2

Oil and Gas in the South China Sea Region[2]

	Proven Oil Reserves (Billion Barrels)	Proven Gas Reserves (Trillion Cubic Feet)	Oil Production (Barrels/Day)	Gas Production (Billion Cubic Feet)
Brunei	1.4	13.8	173,603	315
Cambodia	0	0	0	0
China*	1 (est.)	3.5 (est.)	273,000	141
Indonesia*	0.2 (est.)	29.7 (est.)	215,000	0
Malaysia	3.9	81.7	809,945	1,437
Philippines	0.3	2.8	3,000	<1
Singapore	0	0	0	0
Taiwan	<0.01	2.7	1,300	33
Thailand	0.3	12.5	140,712	565
Vietnam	0.6	6.8	240,000	19
Total	7.7 (est.)	153.5 (est.)	1,856,560	2,510

*Only the regions around the South China Sea are included. Note: There are no proven reserves for the Spratly and Paracel Islands.

Proven oil and natural gas reserves are as of 1/1/2000 (except China and Indonesia South China Sea regions, where data is as of 1998).

Oil production is the 1999 average (except China and Indonesia South China Sea regions, where data is as of 1998). Oil supply includes crude oil, natural gas plant liquids, and other liquids.

Natural gas production is the 1998 average.

"est." means estimated.

The fact that surrounding areas are rich in oil deposits has led to the speculation that the Spratly Islands could be an untapped oil-bearing province located near some of the world's largest future energy-consuming countries. Speculation that the Spratly Islands could have great strategic value has fueled disputes over ownership. In fact, there is little evidence outside of Chinese claims to support the view that the region contains extensive oil resources. Due to a lack of exploratory drilling, there are no proven oil reserve estimates for the Spratly or Paracel Islands, and no commercial oil or gas has been discovered there.

Resource estimates for this region that have been reported in the Chinese press or attributed to Chinese officials vary greatly. Optimistic Chinese estimates of the South China Sea region's oil potential, however, have helped

Table 3.3

Oil and Gas in the South China Sea—Comparison with Other Regions[3]

	Proven Oil Reserves	Proven Gas Reserves (Trillion Cubic Feet)	Oil Production (Million Barrels/Day)	Gas Production (Trillion Cubic Feet/Year)
Caspian Sea Region	16.0–32.5	236–337	1.1	2.9
Gulf of Mexico (U.S.)	2.9	27.3	1.5	5.1
North Sea Region	17.5	146.0	6.2	8.6
Persian Gulf	663.8	1,695.2	20.0	6.1
South China Sea	7.7 (est.)	153.5 (est.)	1.9	2.5
West Africa/ Gulf of Guinea*	32.6	137	3.8	0.2

*Region stretching from Côte d'Ivoire (Ivory Coast) to Angola.

Proven reserves as of 1/1/2000 (except for Gulf of Mexico, where data is of 1/1/99).

Oil production is the 1999 average (except for Gulf of Mexico where data is as of 1998). Oil supply includes crude oil, natural gas plant liquids, and other liquids.

Natural gas production is the 1998 average (except for Gulf of Mexico where data is as of 1998).

"est." means estimated.

encourage interest in the area, with one report suggesting that the Spratly Islands region could become another Persian Gulf. One of the more moderate Chinese estimates suggested that potential oil resources (not proven reserves) of the Spratly and Paracel Islands could be as high as 105 billion barrels of oil, and another suggested that the total for the South China Sea could be as high as 213 billion barrels. A common rule-of-thumb for such frontier areas as the Spratly Islands is that perhaps 10 percent of the potential resources can be economically recovered. Using this rule, the Chinese estimates imply potential production levels for the Spratly Islands of 1.4–1.9 million barrels/day (at reserve/production ratios of 15 and 20)—comparable to 1999 oil production for the entire South China Sea region. The highest Chinese reserves estimate implies that production levels are twice as high as this.

China's optimistic view of the South China Sea's hydrocarbon potential is not shared by most non-Chinese analysts. A 1993/1994 estimate by the U.S. Geological Survey[4], for example, estimated the sum total of discovered reserves and undiscovered resources in the offshore basins of the South China Sea at 28 billion barrels. The most optimistic western estimates place total oil resources (not proven reserves) in the Spratly Islands at 10 billion barrels. Using the same rule-of-thumb, these reserves could yield a peak oil production level for the Spratly Islands of 137,000–183,000 barrels per day—the same order of magnitude as current production levels in Brunei or Vietnam.

NATURAL GAS

Though sometimes overlooked, natural gas might be the most abundant hydrocarbon resource in the South China Sea. Most of the hydrocarbon fields explored in the South China Sea regions of Brunei, Indonesia, Malaysia, Thailand, Vietnam, and the Philippines contain natural gas, not oil. Estimates by the U.S. Geological Survey and others indicate that about 60–70 percent of the region's hydrocarbon resources are gas.

At the same time, natural gas usage among developing East Asian countries (excluding India) is expected to rise by over seven percent annually on average over the next two decades—faster than any other fuel—with almost half of this increase coming from China. If this growth rate is maintained, demand will exceed 20 trillion cubic feet (Tcf) per year—quadruple current consumption levels—by 2020. Gas consumption could increase even faster if additional infrastructure is built. Proposals have been made to link the gas producing and consuming regions of the Pacific Rim region of Asia by pipeline, with the South China Sea geographically central to these regions.

Malaysia is not only the biggest oil producer in the region, in addition it also is the dominant natural gas producer as well, and until recently has been the primary source of growth in regional gas production. The development of natural gas resources outside of Malaysia has been hampered by the lack of infrastructure. Despite this constraint, natural gas exploration activity elsewhere in the region has been increasing. Much of this new activity has occurred in the Gulf of Thailand, offshore China, in Indonesia around the Natuna Islands, and in Vietnam in the Nam Con Son basin southeast of Vietnam.

As with oil, estimates of the South China Sea's natural gas resources vary widely, and resource estimates for this region that have been reported in the Chinese press or attributed to Chinese officials show a large variation. One of the more optimistic Chinese reports estimates that there are 225 billion barrels oil equivalent of hydrocarbons in the Spratly Islands alone. If 70 percent of these hydrocarbons are gas as some studies suggest, total gas resources (as opposed to proven reserves) would be almost 900 Tcf. If the rule of thumb for frontier areas were applied to these resource levels, the Chinese estimates would imply potential production levels for the Spratly Islands of almost 1.8–2.2 Tcf annually (at common natural gas reserve/production ratios in the region of 40–50). The entire South China Sea has been estimated by the Chinese to contain more than 2,000 Tcf of natural gas resources. These totals do not include the unknown potential of the natural gas hydrates discovered off of the Paracel Islands.

This optimistic view of the South China Sea's natural gas potential is not shared by most non-Chinese analysts. The U.S. Geological Survey has estimated that the sum total of discovered natural gas reserves and undiscovered natural gas resources in the offshore basins of the South China Sea is about 266 Tcf. The most optimistic western estimates place total natural gas resources (not proven reserves) in the Spratly Islands at 35 Tcf. Using the same rules of thumb, these reserves could yield peak natural gas production levels for the Spratly Islands of 70–88 billion cubic feet per year—a little less than current production levels in offshore China. These totals are far less than current Malaysian production of over 1,400 billion cubic feet per year.

SHIPPING[5]

In early 1995, then U.S. Secretary of State Warren Christopher issued a warning to the nations quarreling over the Spratly Islands that the United States would not accept the disruption of trade passing through the South China Sea.[6] The warning came because the United States (as well as other trading nations) has vested economic interests to protect in the region. Blockage of shipping lanes, or Sea Lines of Communications (SLOCs), in the South China Sea could immediately disrupt shipping worldwide, hurting international trade and world economies. The sheer volume of merchant shipping transiting the South China Sea gives the region global significance.

More than half of the world's annual merchant fleet tonnage passes through the Straits of Malacca, Sunda, and Lombok, with the majority continuing on into the South China Sea. Oil flows past the Spratly Islands in the South China Sea are estimated to reach 9.5 million barrels per day in 2000, three times greater than through the Suez Canal/Sumed Pipeline, and fifteen times greater than oil flows through the Panama Canal. Virtually all shipping that passes through the Malacca and Sunda Straits must pass near the Spratly Islands. The other major shipping lane in Southeast Asia uses the Lombok and Makassar Straits, and continues into the Philippine Sea. Except for north-south traffic from Australia, it is not used as extensively as the Strait of Malacca and the South China Sea, because for most voyages it represents a longer voyage by several hundred miles.

Shipping (by tonnage) in the South China Sea is dominated by raw materials en route to East Asian countries. Tonnage via Malacca and the Spratly Islands is dominated by liquid bulk such as crude oil and liquefied natural gas (LNG), with dry bulk (mostly coal and iron ore) in second place. Nearly two-thirds of the tonnage passing through the Strait of Malacca, and half of the volume passing the Spratly Islands, is crude oil from the Persian Gulf. Oil flows through the Strait of Malacca rose to 9.5 million barrels/day in 1997, and rising Asian oil demand could almost double these flows over the next two decades.

Northeast Asian nations are heavily dependent upon energy shipments through the South China Sea. More than 80 percent of the crude oil supplies for Japan, South Korea, and Taiwan flow through the South China Sea from the Middle East, Africa, and nations such as Indonesia and Malaysia. Liquefied Natural Gas (LNG) and coal from Indonesia, South Africa, and Vietnam also are shipped via this route. As a result, about two-thirds of South Korean energy supplies, and almost 60 percent of Japan and Taiwan's energy supplies flow through the South China Sea.

LIQUEFIED NATURAL GAS (LNG)

The bulk of the world's LNG trade passes through the South China Sea, and LNG shipments through this sea to northeast Asian markets constituted two-thirds of the world's LNG trade in 1998. Japan is by far the world's largest consumer of LNG, and shipments to Japan accounted for about three quarters of the trade through the South China Sea in 1998, with Japan dependent upon LNG for over 12 percent of its total energy supplies. Shipments to South Korea (the world's second largest consumer of LNG) and Taiwan (the world's fifth largest consumer of LNG) accounted for the remaining shipments through the South China Sea.

Most of this LNG is supplied by Indonesia and South China Sea producers, with Middle Eastern shipments from Abu Dhabi and Qatar also passing through the South China Sea. The South China Sea region also is an important supplier of LNG, with Brunei and Malaysia accounting for about a quarter of the total world LNG production in 1998.

PIRACY

The large volume of shipping in the South China Sea/Strait of Malacca has created opportunities for attacks on merchant shipping. In 1998, 100 attacks—almost half of the world's reported cases of piracy—occurred in this area.[7] Oil product tankers have been among the ships attacked, with 1999 targets including the Thai tanker MV Tenyu off the coast of Malaysia, the Singapore-owned tanker Petro Ranger sailing from Ho Chi Minh City to Singapore, the Indonesian tanker MT Atlanta in the Riau Straits off Sumatra, and the Honduran tanker MT 1 off the eastern coast of Malaysia. The three littoral states of the Malacca Strait (Indonesia, Malaysia, and Singapore) implemented a coordinated patrol and other counter-measures in the region in 1992 that resulted in a decline in the number of attacks in the Strait from 50 in 1991 to 6 in 1998. China also has suggested that joint patrols be conducted in the Spratly Islands to combat piracy and drug smuggling.

WHAT IF THE SOUTH CHINA SEA SHIPPING LANES CLOSED?

Sea lane closure is not in the economic interests of the South China Sea nations. If fighting did break out in the Spratlys, however, combatants would have to cross the second busiest international sea lane in the world to reach the battle zone. It is not necessary, however, to barricade the shipping lanes physically or militarily to achieve the effects of closure. If a threat to shipping is credible, vessels are likely to use other routes. In addition, if an area is declared to be a war zone, maritime insurers will either refuse to insure or might increase the rates of their policies. These financial risks may render a shipping lane sufficiently unattractive to lead to virtual closure, and cause shipping to be diverted to other routes.

The Lombok and Makassar Straits offer a possible alternative route around the South China Sea. This is a longer voyage by several hundred miles, so that nearly half of the world's fleet would be required to sail farther distances, generating a substantial increase in the requirement for vessel capacity. On average, about 15 percent of oil tanker capacity is in commission but idle at any time waiting for charter. However, the volume of traffic through the South China Sea is so large that if vessels rerouted through the Lombok-Makassar route, over 20 percent more tanker volume would be needed to transport this oil.

Although the world's tanker fleet could compensate somewhat, all excess capacity would be absorbed, and worldwide shipping rates could soar, with the oil tanker markets hit hardest. The experience with the Suez Canal closure in 1956 indicates that such a disruption might temporarily increase freight rates by as much as 500 percent *worldwide*, even though the disruption would be localized.[8] Oil prices would temporarily increase as well to reflect the increased shipping rates; shipping and insurance costs currently account for $1–2 U.S. per barrel of imported American oil. A dispute over oil in the South China Sea would have the unintended effect of disrupting energy shipments and prices worldwide.

CONCLUSIONS

The potential of the South China Sea to contain vast amounts of oil has created a rush by littoral nations to carve out territorial niches for themselves. As a result, oil is often viewed as a catalyst for conflict in both the South China Sea and the Spratly Islands in much the same way as it is in the Persian Gulf. This assumption may be erroneous, however, because no commercial oil has been discovered in the central area of the South China Sea or the Spratly Islands, and no proven oil reserves have been found in this region. The oil resources in the Spratly Islands and the central South China Sea are likely to be in small quantities and difficult to extract. Natural gas, rather than oil, is more likely to be present in sizable amounts in the South China Sea. This fact, combined with fast-rising natural gas utilization rates in the Asia-Pacific region, may make the South China Sea's natural gas, not oil, the strategic commodity worth fighting for in the future.

Nations which rely on seaborne trade have a vested interest in preserving the stability of the Southeast Asian shipping lanes because a localized disruption in shipping will cause an impact on economic markets worldwide. The factor that converts a local maritime concern—a closure of a shipping lane—into a global economic event is the large volume of shipping and world trade that transits the South China Sea. While a war might start over a suspected oil field in the South China Sea, it is shipping that is the strategic issue on a global basis.[9]

NOTES

1. UNCLOS adopted December 10, 1982 and the subsequent Agreement Relating to the Implementation of Part XI of the UNCLOS of December 10, 1982, adopted July 28, 1994, United Nations, Division for Ocean Affairs and the Law of the Sea, Office of Legal Affairs, New York. These reports are available at the United Nations Web site at www.un.org/Depts/los/losconv1.htm.

2. Oil and natural gas reserve estimates in Table 3.2 (except for China and Indonesia) are from the *Oil and Gas Journal*, December 20, 1999. Chinese reserves in the South China Region are U.S. Energy Information Administration estimates. Indonesian reserves in the South China Sea Region were compiled from the *Petroleum Report Indonesia*, produced by the Embassy of the United States in Jakarta. Oil production estimates are from the U.S. Energy Information Administration. Natural gas production estimates are from the *International Energy Annual*, U.S. Energy Information Administration.

3. Oil and natural gas reserve estimates in Table 3.3 (except for the Caspian Sea and the South China Sea) are from the *Oil and Gas Journal* (December 20, 1999). Caspian reserves are U.S. Energy Information Administration estimates. South China Sea reserves are from Table 2.1 of this report (see note 7). Oil production estimates

are from the U.S. Energy Information Administration. Natural gas production estimates are from the *International Energy Annual*, U.S. Energy Information Administration.

4. Charles D. Masters, Emil D. Attanasi, and David H. Root, "World Petroleum Assessment and Analysis" in *Proceedings of the 14th World Petroleum Congress* (John Wiley and Sons, 1994), 529–541. This and other U.S. Geological Survey reports on world energy resources are available at http://energy.er.usgs.gov/products/papers/World_oil/.

5. Much of the shipping data and analysis were abstracted from *Chokepoints: Maritime Economic Concerns in Southeast Asia* by John Noer and David Gregory, a collaborative effort by the Center for Naval Analyses and the Institute for National Strategic Studies (1996). Data on LNG shipments in the region and the dependency of countries for energy supplies flowing through the South China Sea are from the U.S. Energy Information Administration.

6. U.S. Policy on freedom of navigation in the region is also outlined in the U.S. Department of State Daily Press Briefing (Wednesday, May 10, 1995), available at http://dosfan.lib.uic.edu/ERC/briefing/daily_briefings/1995/9505/950510db.html.

7. *Piracy and Armed Robbery Against Ships 1999*, International Chamber of Commerce, International Maritime Bureau, (Essex, U.K.), available at http://www.iccwbo.org/home/news_archives/2000/piracy_report. Additional information on oil piracy in the South China Sea is available in "Oil Piracy Poses Growing Menace to Tanker Traffic in South China Sea," *Oil and Gas Journal* (October 18, 1999).

8. John Noer and David Gregory, *Chokepoints: Maritime Economic Concerns in Southeast Asia* (Center for Naval Analyses and the Institute for National Strategic Studies, 1996), 40–43.

9. The U.S. Energy Information Administration updates its assessment of the energy situation in the South China Sea annually. This report is available at http://www.eia.doe.gov/emeu/cabs/schin.

4

Security Risks of a South China Sea Conflict

David G. Wiencek and John C. Baker

INTRODUCTION

The South China Sea is home to several flashpoints that present major security risks to regional stability. The nations that are contesting this disputed maritime territory continue to jockey for position and influence. Small military outposts, sometimes put in place under the guise of fishing shelters or scientific research stations with a purported civilian orientation, are being constructed with increasing frequency. The number of island occupations sharply increased in the 1980s and continued in the 1990s, and this trend shows no sign of abating in the early part of the twenty-first century.

The drive to occupy territory has given rise to incidents of violence. In 1974, China displaced Vietnam from the Paracels, in the northern reaches of the South China Sea.[1] In 1988, Vietnam and China fought the bloodiest battle to date in the South China Sea in a dispute over the Fiery Cross Reef in the Spratlys, which resulted in the loss of three Vietnamese ships and over 70 Vietnamese sailors killed or missing; others were captured. In recent years, we have seen a range of provocations take place among the different claimants, from overflying a contested island, to ship collisions, to the arrest and harassment of fishing vessels and fishermen. The close proximity of the disputants—Vietnamese-claimed Southwest Cay, for example, lies within eyeshot of Philippine-claimed Northeast Cay in the Spratlys—also heightens the risk of future incidents and military clashes.

Against this background and looming large in the near future is China's military buildup with its emphasis on power projection and blue water naval capabilities. China is clearly pursuing a strategy of expanding its military sphere of influence in the area to include strategic waypoints in the Paracel Islands (particularly Woody Island), down through the Spratlys, as a way of enforcing its expansive claim to the entire South China Sea. (For geographic perspective, the controversial Mischief Reef, which is claimed by China, Taiwan, Vietnam, and the Philippines, is located 700 miles from China's Hainan Island and over 1,000 miles from the Chinese mainland.)

This orientation puts China on a collision course with the other claimants, and, more importantly, with the United States and its allies, who are committed to ensuring freedom of navigation and flight paths in the vital commercial and strategic passages of the South China Sea. The April 1, 2001, collision between a Chinese F-8/J-8-II fighter and a U.S. EP-3 reconnaissance plane over the South China Sea illustrates the potential for inadvertent conflicts arising in the region. Any military clash in the area, therefore, could very well escalate and draw in others in an effort to keep the sea lanes and air corridors from being disrupted.

This chapter reviews the security risks arising from the South China Sea disputes by examining: 1) how a conflict might arise; 2) regional security perceptions and interests; 3) recent military activities and maneuverings among the various claimants; and 4) how China's on-going, long-term military expansion threatens to exacerbate regional tensions. It concludes by assessing the implications for the regional states and others, including the United States, of a conflict in the South China Sea.

CONFLICT POTENTIAL IN THE SOUTH CHINA SEA

As discussed throughout this book, there are a number of indicators that point to future instability in the South China Sea. These include the following:

- Intractable sovereignty disputes
- The quest for valuable natural resources, including oil and natural gas
- Deployed military forces
- A track record of previous clashes
- Continuous diplomatic jousting
- High potential for provocative acts and miscalculation
- Weak regional mechanisms for preventing conflict

This backdrop helps explain the existing pattern of behavior that is making conciliation and negotiation difficult. The disputants, therefore, continue to engage in conflictive actions and clashes.

Although scenarios could vary, there are several main triggers or situations that could lead to armed confrontation.[2] These include the following:

- *Creeping Occupation.* Each claimant has pursued a policy of "effective occupation" of key islands and reefs among the Spratly Islands to reinforce their sovereignty claims. In some cases, this has entailed constructing outposts, such as the Chinese facilities at Mischief Reef, upon maritime features because land was unavailable.[3] The cumulative effects of this type of island/reef grabbing activity, sometimes undertaken by stealth and surprise, could, over time, lead a claimant to come to believe that its position is seriously threatened and that an armed response is a necessary option to redress the situation.
- *Exploration or Exploitation Activity.* Unilateral steps to explore or exploit the hydrocarbon resources of the South China Sea also can be a major catalyst for conflict because such actions directly raise the sovereignty issue. China has been quite sensitive to the efforts of other claimants, such as Vietnam and the Philippines, to grant oil and natural gas concessions to commercial firms. Beijing or another disputant could thus seek to oust the drilling fleet/rigs of another claimant by military force, thereby putting foreign commercial interests and oil or gas company personnel at risk, as well.
- *Aggressive Patrolling.* A more routine source of incidents in the South China Sea involves aggressive maritime patrols using naval ships and military aircraft that run the risk that an incident at sea will escalate into an armed clash between the claimants.[4] These incidents usually involve one claimant seizing another country's fishing boats and detaining the crews to demonstrate a willingness to enforce its claims. Along these lines, sources in Taiwan report that between 1992 and 1996, 134 Taiwanese fishing boats and trawlers were harassed, inspected, detained, rammed, or fired upon by Chinese naval vessels in the northern part of the South China Sea.[5]
- *Armed Displacement.* The claimants could resort to military force to press their territorial claims even though the likelihood of armed conflict is normally not very high given the serious diplomatic consequences of fighting and the limited capabilities of claimants for projecting their military power at this time. As noted, however, some instances of military intimidation have already occurred, and the possibility remains that one claimant could initiate a direct military attack against the outpost or forces of another. For its part, China has emplaced anti-aircraft and naval guns on many of its facilities, such as Mischief Reef, an overt move to fortify its positions and repel possible attackers when required. It could also be argued that China's aggressive interception of the U.S. EP-3 aircraft in April 2001, was an act of intimidation designed to limit U.S. reconnaissance of Chinese military activities in and around the South China Sea. The Chinese fighter reportedly closed to within feet of the U.S. aircraft and then lost control, crashed, and damaged the EP-3, forcing it to make an emergency landing on Hainan Island.

Finally, troop deployments in the Spratlys are difficult to ascertain but believed to be small. One widely-quoted open-source estimate made in the early 1990s (as depicted in Table 1.1) put the total number of troops deployed by all the claimants at about 1,500.[6] But today, the numbers remain uncertain. A classified U.S. Government report obtained by the Kyodo news agency placed the number of Chinese troops in the Spratlys in 1999 at 325.[7] Taiwan was generally believed to station approximately 110 Marines on Itu Aba/Taiping Dao, its only Spratly possession. However, the government decided in late 1999 to withdraw the Marines because it felt incapable of holding the islands or resupplying the garrison there in the event of war. Coast guard personnel instead were given the responsibility for security at Itu Aba. At that time, reports from Taiwan stated that around 1,000 Taiwanese soldiers were stationed both on Itu Aba and on Pratas Island, a separate (disputed) feature southwest of Taiwan and well outside the Spratly area.[8] The Philippines may deploy far fewer troops on its Spratly possessions due to budget constraints. Whatever the actual numbers may be at this time, any conflict in the Spratlys would only marginally depend on troops already deployed (who are in any event likely in a generally low state of military readiness). Rather, future military battles in the Spratlys will be fought mainly by those front-line air and sea forces that can be marshaled from outside as the need arises and introduced into the theater of conflict.

REGIONAL SECURITY PERCEPTIONS AND INTERESTS

The following represents a survey of the differing security perceptions and interests of the claimants and others affected by the Spratly Islands disputes.

China

The People's Republic of China (PRC) is the most important and influential regional actor. However, it has made a vast claim to the entire South China Sea, thus making confidence-building measures (CBMs) and diplomacy highly problematic and dependent on satisfying Beijing's interests. In 1992, China adopted a Law on Territorial Waters and Their Contiguous Areas, which reiterated its sovereignty claims in the South China Sea. Importantly, in this directive, Beijing strongly implied it would use military means to enforce innocent passage in the territorial seas around its claimed islands.[9] The legal basis of China's claims has been backed up by a seemingly calculated set of island occupations over time, thus enabling China to establish a permanent and continuous military presence in the Spratlys. A number of permanent fortresses have been constructed, including those on the following reefs: Subi, Fiery Cross, Mischief, Johnson South, and Chigua. These outposts have some self-defense capability, including anti-aircraft and naval guns, and landing pads that can accommodate military helicopters. Reports also indicate the presence of Silkworm anti-ship cruise missile installations on Woody Island in the Paracels in the northern region of the South China Sea.[10] The Silkworm missile has a range of 95 km/59 miles.

Furthermore, China has linked its island occupations with an expanding military strategy and force buildup that is designed to project power to the far reaches of the South China Sea and beyond. China is thus positioning itself to exert control—in time—of the region's vital sea lanes and airspace. It views the other claimants as challenging this predominant position. It also perceives attempts by outside nations, including the United States, as threatening and unwanted interference.[11]

Malaysia

Malaysia holds four of six reefs (Ardasier, Mariveles, Swallow, and a portion of Investigator Shoal) in its claimed area in the Spratlys. In 1991, Malaysia began constructing a resort and airstrip on Swallow Reef. But after years of adopting a relatively low-key posture in the Spratlys, Malaysia secured an outpost on Investigator Shoal in mid-1999 and may have occupied two additional nearby reefs. (Kuala Lumpur reportedly fabricated a two-story concrete building in Penang, in peninsular Malaysia, and had it towed to Investigator Shoal, some 700—800 nautical miles away.[12]) Prime Minister Mahathir described the new construction as follows: "We have built on our own zone and not outside the zone for climate research and marine life studies and also to prevent ship collisions."[13]

However, these moves prompted complaints from the Philippines, which also claims Investigator Shoal as Pawikan Shoal (its Filipino name). In October 1999, Malaysian and Philippine aircraft reportedly came into contact without incident near Investigator/Pawikan Shoal (discussed later).

It is unclear why Malaysia moved to raise its profile in the Spratlys. It could have been in response to events in the Philippines, where Washington and Manila upgraded defense ties in 1999, although Manila still lacks the military capability to enforce its claims. There also has been some speculation that Malaysian authorities may have been upset with Filipino counterparts over the latter's support for Deputy Prime Minister Anwar Ibrahim, who was ousted from power in late 1998. Thus, Malaysia's moves in the Spratlys could have been a way of demonstrating Malaysia's unhappiness with Manila.

Another explanation could be that Malaysia reached a tacit understanding with China to try and move the dispute away from the multilateral realm, into bilateral negotiations between Beijing and Kuala Lumpur. In 1999, Malaysia appeared to upgrade its relations with China. In January 1999, the People's Liberation Army (PLA) Chief of Staff visited Kuala Lumpur. Then, in May, Malaysia's Foreign Minister visited Beijing and endorsed a bilateral "Joint Statement on Framework for Future Bilateral Cooperation." Among other things, the Framework noted that the two sides would work to improve defense cooperation, including the exchange of information and intelligence, reciprocal personnel and ship visits, and training. Regarding the South China Sea disputes, the Framework indicates that the matter should be solved through "bilateral friendly consultations."[14] Malaysia's apparent willingness to resolve this issue through bilateral diplomacy is an important departure from previous ASEAN positions. In the past, the Spratly dispute had mainly been one of China versus ASEAN, and the smaller Southeast Asian nations were able to unite in an effective diplomatic posture against their larger neighbor's maneuverings. It now appears, however, those calculations may have changed and Malaysia may have reached some form of agreement with China about the future of this important issue—an understanding that effectively undercuts ASEAN solidarity. Meanwhile, Malaysia is continuing to procure the advanced air and naval assets that will enable it to protect its territorial claims and offshore resources, including oil and gas reserves.

The Philippines

In recent years, the Philippines has been an active participant in the diplomatic and military maneuverings in the Spratly area, viewing the actions of China, in particular, as threatening to its national security. In what we have called the Mischief Reef I and II incidents of 1995 and 1998, Manila perceived Chinese actions as threatening to dominate by force Philippine-claimed areas. Furthermore, there also have been a number of low-level security-related incidents between the two sides over the past several years, and this pattern looks likely to continue. However, there is little that the Philippines can do militarily at present to thwart Chinese probing and island occupations in the Spratlys or the contested fishing waters around Scarborough Shoal. The Armed Forces of the Philippines (AFP) sorely lacks air and naval capabilities, and is in the process of a multi-year defense modernization and equipment upgrade effort. However, the Asian financial crisis of 1997–1998 severely setback AFP military acquisition plans. As one Philippine national security official described the situation: "At the moment China is like a dog peeing everywhere to mark out its territory ... we are on the edge of their claim and we are the weak link, so we suffer."[15] This situation has been exacerbated by internal political turmoil and security problems, including the continuing Muslim insurgency in Mindanao, and a persistent national political crisis that resulted in the President Joseph Estrada's removal from office in January 2001 on charges of corruption. Estrada's Vice President, Gloria Macapagal-Arroyo, assumed the presidency in the peaceful change of power.

Policymakers in Manila have resorted to public diplomacy as one of the few tools available to help draw attention to threats to the country's interests; they have continually voiced their concerns in the international media and at many regional and international diplomatic meetings.

The Philippines also has sought, unsuccessfully, to call on its long-time treaty ally, the United States, to offer support for its claims in the Spratlys. However, Washington has maintained a position of neutrality regarding the conflicting claims in the South China Sea, and it has effectively told Manila that a position of ambiguity is important for maintaining deterrence in the region. To date, Washington has deferred from endorsing Manila's interpretation that the Mutual Defense Treaty between the two countries obligates the United States to come to the aid of the Philippines should Philippine interests be threatened in the South China Sea. This posture essentially leaves Manila exposed to further encroachments on its territorial claims. A more positive development occurred in May 1999, however, when the Philippine Senate approved the Visiting Forces Agreement (VFA), a measure that updated previous agreements governing the legal status of U.S. forces. The VFA provided a basis for the U.S. military to begin operating again (in the aftermath of the U.S. departure from Subic Bay in 1991) in conjunction with Philippine forces for training and exercises. Beyond its provisions, the VFA signaled that the two longtime allies are in fact prepared to go forward working to rebuild deterrence in Southeast Asia, and it demonstrated a commitment by both sides to work to prevent Beijing from destabilizing the region.

Taiwan

Taiwan views its interests in the South China Sea largely through the prism of its broader dispute with the PRC. Understandably, Taipei must accord priority to the defense of Taiwan and the military balance across the Taiwan Strait. Offshore islands, including Itu Aba/Taiping Dao in the Spratlys, must be accorded secondary priority from a defense and security standpoint.

Taiwan's claim to the Spratlys replicates that of the PRC. Taiwan was the first government to establish a physical presence on a Spratly feature, Itu Aba, which it has occupied since 1956. By the late 1990s, however, Taipei came to recognize that the defense of Itu Aba was impractical, and it withdrew its contingent of Marines stationed there and turned security over to the Coast Guard. This move came after a period of continuous high tensions between Taiwan and the PRC. In 1995–1996, Beijing conducted large-scale military maneuvers opposite Taiwan and staged a series of highly provocative missile "tests" in an effort to intimidate Taiwanese authorities and the country's electorate. In addition, there also was concern that China would stage additional threatening actions prior to the presidential election of March 2000. Hence, the move to withdraw the Marines from Itu Aba could have been a means to lessen the attractiveness of the island as a target for Beijing and to diminish the potential negative political consequences that could occur if the PRC takes the island by force.

However, if tensions between the two states persist, the PRC could be tempted to harass or attack Itu Aba. The PRC could defend such an action to the world as an internal matter between it and Taiwan that pertains only to their mutually compatible territorial claims in the South China Sea. Finally, a PRC attack on Itu Aba would avoid the risks associated with military action in the Taiwan Strait.[16]

Taiwan, meanwhile, continues to upgrade its military forces in an attempt to maintain a balance of power across the Taiwan Strait. It also continues to look to Washington, under the Taiwan Relations Act of 1979, as a guarantor of its security should the PRC seek to resolve their dispute by force.

Vietnam

Vietnam asserts a claim to both the Paracel and Spratly islands. It is also engaged in a maritime dispute with China over the Gulf of Tonkin. However, Vietnam and China have undertaken periodic discussions at resolving their land border disputes and begun talks on their conflicting maritime claims.

For Hanoi, enforcing claims in what it calls the Eastern Sea is important from a national defense perspective, broadly helping to ensure the security of its long eastern coastline. The region is also important from an economic security perspective, including access to offshore resources, such as oil.[17]

Vietnam occupies the largest number of features in the Spratlys, now estimated to be 27 islands. It has built a 600 meter runway on Spratly Island itself. It has not been shy about engaging other claimants when it perceives them to be encroaching on its territorial claims, and has tangled in varying security-related incidents with China, Taiwan, and the Philippines.

Hanoi continues to gradually upgrade its air and naval capabilities in an effort to bolster its claims. It has purchased 12 Su-27 Flanker fighters from Russia, and may acquire more of these and/or Su-30 multi-role combat aircraft, which are equipped for air refueling and long distance missions. In 1998, Vietnam also acquired two Sang-o class submarines from North Korea. These submarines have a range of 2,700 nm, and could be used for mine laying. There also is a report that the submarines are being equipped with an unspecified missile capability.[18] These submarines make Vietnam only the second claimant after China to possess an underwater strike capability, albeit a limited one at present.

Indonesia

Indonesia is not a direct party to the Spratly Island disputes. Nevertheless, Jakarta has taken an active role over the years as an honest broker in attempting to help mediate the dispute via the series of informal workshops it has sponsored among the disputants. This role, however, has substantially diminished since the onset of the 1997 Asian financial crisis, which sparked a national political crisis that continues to the present. Indonesia confronts serious political, security, and economic challenges arising from the power shift of 1998, when long-serving President Suharto was forced to step down by a burgeoning democracy movement in the aftermath of financial crisis. Suharto's departure eventually led to the country's first free elections in over 40 years and the election of a new president, Abdurrahman Wahid, who was later ousted in 2001. Among the critical and immediate security concerns facing Indonesia's leaders is separatist violence in such key locations as Ambon and Aceh. These internal strains, not potential offshore territorial disputes, are likely to remain the top priority of Indonesian policymakers for the near future.

Still, Jakarta has a major strategic interest in preserving its claims around Natuna Island, which lies well to the south of the Spratlys. Indonesia's ownership of the gas-rich fields off Natuna was undisputed until China released an official map with unclear maritime boundaries indicating that Chinese-claimed waters in the South China Sea may extend into the waters around the Natuna Islands. Indonesia responded by choosing the Natuna Islands region as the site of its last major multiservice training exercise in 1996. Indonesia, like many others in Southeast Asia, remains concerned over Beijing's intentions.

Australia

Australia is economically dependent on open sea lanes in Southeast Asia. It has been estimated that five percent of all its exports (as a percentage of GDP) and seven percent of its imports pass through these sea lanes.[19] Australian security interests thus extend to Southeast Asia and the South China Sea region. Australia has been particularly supportive of ASEAN diplomatic efforts. It also maintains defense ties with the United States and works closely on security issues with its U.S. ally. At the same time, it is engaged with regional partners, such as Indonesia, as well as with Malaysia and Singapore under the Five Power Defence Arrangement.

Japan

Japan depends heavily on the flow of free trade through the critical maritime passages of the South China Sea. Almost 60 percent of its total energy supplies are shipped through this region, while some 40 percent of the country's total imports and exports have been estimated to transit these sea lanes.[20] Thus, maintaining freedom of navigation through the South China Sea must be seen as a strategic priority for Tokyo. Japan has found it difficult to play a more active role on this issue, however, due to historical constraints on its armed forces. Upgraded security ties with the United States under a set of new Defense Guidelines issued by the two allies in 1997 clarified Tokyo's role in responding to situations in areas surrounding Japan, and suggest that Japanese forces could become more actively involved in regional security contingencies in the future.

India

India is an emerging Asian power that eyes China as a major future political competitor. It has indicated that it, too, was interested in developments in the South China Sea. In 2000, India held joint military (counter-piracy) exercises with Vietnam.[21] Indian naval vessels had plans to participate in the exercises and also make port visits in South Korea, Japan, and in China itself.[22]

South Korea

South Korea's primary security interest is naturally the military threat posed by North Korea, and secondarily the general balance of power in Northeast Asia. Yet Seoul, too, must be concerned with developments in the South China Sea due to the high dependency of its economy on trade that flows through this region. As our colleague Erik Kreil pointed out in Chapter 3, some two-thirds of South Korean energy supplies pass through the South China Sea.

Thailand

Thailand is not a party to the Spratly Island dispute, but has been concerned for some time over this issue. It fears that a military clash could disrupt its trade routes to East Asia and the United States. It is also concerned that fighting could spill over to the Gulf of Thailand and pose a threat to its sea and land frontiers. Bangkok also remains apprehensive of Chinese intentions. Due to the Asian financial crisis, however, Thailand was forced to sharply curtail its military modernization program (which included Southeast Asia's first sea-based, air power system), and has now turned its primary defense planning focus to economic-based security issues and border concerns with Burma.

United States

A military confrontation between China and one or more of the ASEAN nations would place the United States in a problematic position.

Although expressing concern over the risk of conflict and the need to maintain freedom of navigation, official U.S. policy can be described as one of "active neutrality." While avoiding any position on the legal merits of the various sovereignty claims in the South China Sea, Washington consistently urges the countries involved to find peaceful means for resolving their disagreements. But outside of encouraging the work of the ASEAN Regional Forum (ARF) and the Indonesian-sponsored "track two" workshops, American officials are particularly circumspect in discussing what diplomatic or military role the United States might assume in a regional conflict sparked by the Spratly Islands disputes.

Nonetheless, a serious military clash involving China and an ASEAN nation, such as the Philippines, could pose a policy-making dilemma for U.S. officials. On the one hand, if the United States exhibits no show of military force, then it might appear to be acquiescing in China's use of military force to resolve the territorial disputes surrounding the Spratly Islands. On the other hand, U.S. military deployments or active support for an ASEAN country would likely have major spillover effects on U.S. diplomatic and trade relations with China. Hence, to avoid this policy dilemma Washington has a strong interest in supporting efforts that seek to avoid or contain potential regional conflicts arising from the Spratly Islands disputes. As such, the United States should develop more robust policy guidelines, emphasizing that it will not tolerate an attempt by any nation to use force to resolve the disputes or to disrupt the vital sea lanes of Southeast Asia. It also should continue to show the flag in the South China Sea in support of the principle of freedom of navigation, as it did in 1998 when it sent an aircraft carrier near the Spratlys.

ASEAN Regional Forum

High expectations have long existed for the ARF assuming a more proactive role in helping to contain the risk of armed conflict in the South China Sea, and particularly the multinational disputes surrounding the Spratlys. Although the ARF has made recurring efforts to promote a Code of Conduct for the South China Sea that sets forth guidelines for maritime activities, as well as discourages new occupations of Spratly locations, these discussions have yet to yield any tangible results.[23] This lack of progress is partly due to the fact that some ASEAN member states, such as Malaysia, have opted to establish new outposts in the Spratlys, while China has traditionally preferred to engage in bilateral diplomacy rather than multilateral talks, thus undercutting the development of a unified ASEAN position concerning the Spratlys.

RECENT MILITARY ACTIVITIES AND MANEUVERINGS

The claimants continue to engage in military-related maneuverings, which run various levels of security risks. The potential risk of conflict is best illustrated by the continuing dispute between China and the Philippines over

Mischief Reef, which lies about 135 miles from the Philippine's coastline and falls within the EEZ claimed by Manila. The conflict potential between the Philippines and China substantially increased in early 1995 when Chinese ships arrived with construction personnel and built octagonal structures on stilts at four locations on the reef. Beijing declared these shelters were for Chinese fishing boats operating in the area. However, Manila viewed the alleged "fishermen structures" as Beijing's effort to advance its territorial and maritime claims to the Spratly Islands, and made strong verbal protests against China's bold action.

After several years of relative calm, the risk of conflict significantly increased once more in late 1998. With the region distracted by the Asian financial crisis, Beijing moved again, upgrading and expanding its facilities on Mischief Reef. In what we have called the Mischief Reef II incident, a group of seven Chinese naval and support ships arrived at Mischief Reef in November 1998, and disembarked a large construction party that proceeded with a substantial upgrade of the earlier modest octagonal "fishermen" structures. They constructed two new permanent, multistory structures on concrete platforms, which are manned by Chinese military personnel.

While Chinese officials declared that the renovation was needed to sustain the deteriorating original structures against weather damage, the Philippines quickly charged China with undertaking a creeping invasion of its sovereign territory. Philippine officials stated that Beijing was pursuing a "talk and take" strategy aimed at consolidating its political and military position in the South China Sea at the expense of the Philippines.

Tensions between the two countries continued through the 1990s. In early 1999, their dispute involving the Spratlys was characterized by heated rhetoric, increasingly close aerial overflights by Philippine aircraft of the Chinese construction sites, and two separate incidents at sea where the Philippine Navy chased down and sank Chinese fishing boats that Manila claimed were illegally operating in the waters of its EEZ.[24]

In addition to the Mischief Reef tangle, there have been a number of other worrisome incidents. In May 1999, for example, two Chinese naval ships allegedly pointed their guns at a grounded Philippine supply ship, while in October 1999, Vietnamese troops fired on a Philippine plane during an overflight of Tennent Reef. A Philippine media report indicated that one of the country's planes was flying as low as 120–200 ft over the reef when the Vietnamese opened fire; the plane was not hit.[25]

Around the same time, two Royal Malaysian Air Force Hawk fighter aircraft on a training flight and two Philippine OV-10 Broncos on a reconnaissance mission reportedly came into contact with each other, but without incident, near Investigator (Pawikan) Shoal. This is a location where Malaysia had constructed a new outpost in an area also claimed by the Philippines.[26]

CHINA'S GROWING MILITARY POWER

As a rapidly emerging power, China is increasingly coming to dominate regional security considerations. Beijing's growing political-military reach raises serious concerns relative to the situation in the Spratlys due to China's vast claim to the entire South China Sea area. China appears to be biding its time until its military capabilities, particularly its South Sea Fleet, are sufficient enough to enable it to back up this broad claim through the use of force. Its island-grabbing actions to date, most notably the recent Mischief I and II episodes, foster the perception held by many that China intends to use its chain of outposts in the Paracel and Spratly Islands to dominate the entire area militarily. If this assessment proves accurate, China would be in a position at a future date to possibly control the vital sea lanes and airspace in, around, and through the South China Sea.

China is currently embarked on a major military modernization program designed to position itself as the dominant power in Asia. Indicators of China's intent to acquire the tools for military power projection include new long-range ballistic and cruise missiles, new nuclear submarines and surface warfare combatants, Russian Su-27 and Su-30 fighter aircraft, long-range refueling and command and control aircraft, and an aircraft carrier to be built in the next several years. Additional carriers could be commissioned by 2010 with the intent of building a power projection fleet to be used to enforce China's claims in the Spratlys, as well as in its major political dispute with Taiwan.[27]

These new military tools will enable China to achieve its enunciated goals of moving away from a past coastal defense orientation in its military strategy to a blue-water, power projection force capable of dominating what Chinese strategists term the First Island Chain, which runs from the Spratlys up to Taiwan and the Ryukyus, and then eventually extending its power to reach the Second Island Chain, which runs out to the Marianas, Guam, and Pelau.[28]

China already has established a major outpost on Woody Island in the Paracels. It is nearly 2 km^2 in size and has a 2,600 m/7,800 ft concrete airstrip, which was built out over the surrounding reef. This airstrip is capable of handling all types of PLA aircraft. A recent report suggests that the Chinese have constructed an aircraft storage facility consisting of four concrete-block hangers, each capable of holding two fighter aircraft, and a hardstand that can accommodate 30 additional aircraft. Between 1991 and 1998, 67 PLA fighters were deployed to the island in 14 deployments.[29] Woody Island also has a 350 meter pier with port facilities and docks, power plants, oil tanks, ammunition storage bunkers, and gun emplacements. Woody Island appears to be designed as a staging point to support offensive operations in the Spratlys. A signals intelligence station, meanwhile, has reportedly been established on Rocky Island, just to the north of Woody Island. Rocky Island is one of the highest points in the area, and thus provides good coverage of signal activity in the northern part of the South China Sea.[30]

China also has one J-17C radar on Fiery Cross Reef that provides some early-warning capabilities. Reportedly, the radar is used to monitor Vietnamese air force activity out to a range of 150 nautical miles.[31]

The other Spratly claimants are not in a position at this time to challenge China militarily. Many major Southeast Asian procurement and equipment upgrade programs were severely curtailed during the Asian financial crisis of 1997–1998, which had a negative impact on the economies of the region. China, by contrast, was less affected, and continues to forge ahead with its military modernization programs.

CONCLUSIONS

Security risks in the South China Sea are real and growing. Even a limited military conflict in the region could have detrimental consequences not only for the disputants, but it also could create significant problems for other Asian countries, such as Japan and South Korea. As noted earlier, a regional conflict arising over the Spratly Islands could disrupt the critical sea lines of communication that pass through the South China Sea. Furthermore, a serious confrontation involving China might create a political stumbling block for regional economic cooperation.

The resurgent risk of armed conflict over the Spratly Islands is all the more dangerous because its routine nature conceals an underlying unpredictability. For the most part, this territorial dispute is characterized by diplomatic exchanges, as well as by occasional provocative actions, such as fishing ship incursions or oil explorations in the disputed areas by one claimant that stimulate another one to take a strong diplomatic response or even a show of force. However, as described previously, some events have escalated into more serious showdowns, including the use of deadly force, such as the 1988 clash between China and Vietnam. The April 2001 incident involving the U.S. reconnaissance aircraft and Chinese fighter further highlights the security sensitivities in this vital region and the looming clash of interests between Washington and Beijing.

Hence, the cyclical pattern of conflict in the South China Sea dispute is manifested in continuing low-level incidents punctuated by sporadic flare-ups among the disputants that occasionally involve the use of force. The possibility of a wider regional conflict arising from military miscalculations by one or more sides in a territorial dispute, such as occurred in the Falklands War between Argentina and the United Kingdom, cannot be completely ruled out given the continuing diplomatic deadlock to resolve the South China Sea disagreements. The United States and other countries concerned with the risks of a military conflict erupting over the Spratly Islands need to be proactively involved in pressing the claimant countries to explore ways to mitigate the risk of armed conflict. At the same time, these countries must continue their efforts to address the underlying diplomatic and economic sources of the Spratly Islands disputes.

NOTES

1. In this skirmish, "one American civilian, who had accompanied the Vietnamese as an observer, was beaten and also taken prisoner." As reported by Captain Anton Nugroho, Indonesian Armed Forces, "The Dragon Looks South," *U.S. Naval Institute Proceedings* (March, 2000), 74.

2. This listing is mainly drawn from Ralph A. Cossa, "Summary and Recommendations," in Carolina G. Hernandez and Ralph Cossa, eds., *Security Implications of Conflict in the South China Sea: Perspectives from Asia-Pacific* (Quezon City: the Philippines: Institute for Strategic Development Studies, Inc. and Pacific Forum/CSIS, 1997), 221–222; and Scott Snyder, "Preventing Escalation of Disputes in the South China Sea: An American Perspective," draft paper prepared for "Promoting Trust and Confidence in Southeast Asia: Cooperation and Conflict Avoidance," co-sponsored by the Institute of Strategic and Development Studies, University of the Philippines, and CSIS/Pacific Forum, (Manila, October 16–18, 1997).

3. In fact, the claimants have been persistent and creative in building structures above and extending from the small island features. One Chinese expert, for example, has described Fiery Cross Reef (Yongshu Jiao) as "originally a small piece of rock about a table in size," which has now been built-up to the size of a full, permanent outpost. See Pan Shiying, "The Nansha Islands: A Chinese Point of View," *The Window* (Hong Kong, September 3, 1993), 28. Chinese commentaries even speak of three generations of different island structures: First Generation "sheds" were built of bamboo; Second Generation "octagonal pavilions" were built of iron sheeting; and the Third Generation features "permanent constructions" in the form of three-story white buildings on concrete platforms, which are referred to by the men who man them as "fortresses in the sea" or "sea bastions."

4. A good example of this aggressive patrolling occurred in May 1997 when a boatload of amateur radio hobbyists from various nations, which sailed out of a Chinese port to Scarborough Shoal, landed on some small volcanic rocks located in the Philippine's declared EEZ in the northern portion of the South China Sea to conduct some broadcasts. They were overflown by military reconnaissance aircraft and later challenged by four Philippine naval gunboats because of Manila's concern that their activities gave credence to China's territorial claims. Shortly afterward, a boatload of Philippine legislators traveled to Scarborough Shoal to make a show of planting the Philippines flag and subsequently the Philippines navy detained some Chinese fisher-

men who were operating near the shoal. Andrew Sherry and Rigoberto Tiglao, "Law of the Seize," *Far Eastern Economic Review* (June 12, 1997), 17, 20–21.

5. Chien Chung, "Confidence-Building Measures in the South China Sea," in Hung-mao Tien and Tun-jen Cheng, eds., *The Security Environment in the Asia-Pacific* (Armonk, N.Y.: M.E. Sharpe, Inc., 2000), 273.

6. Zhiguo Gao, "The South China Sea: From Conflict to Cooperation?," *Ocean Development and International Law* 25 (1994), Table 1, 348.

7. See Dario Agnote, "China to Continue Military Buildup in Spratly Islands," *Washington Times* (October 21, 1999), A14.

8. "Tang reiterates reasons for pullout from islands," *China News*, (December 7, 1999).

9. "The Law of the People's Republic of China on its Territorial Waters and Their Contiguous Areas," *Xinhua*, (February 25, 1992), as translated by BBC, FE-1316-C1-1 (February 28, 1992), (see particularly Articles 6, 8, 10, and 14). Also see Bradford L. Thomas and Daniel J. Dzurek, "The Spratly Islands Dispute," *Geopolitics and International Boundaries* (Winter 1996), 306. Following the April 1, 2001, incident between involving the Chinese fighter and the U.S. reconnaissance aircraft, an article carried by the official *Xinhua* news agency, noted that Article 12 of this Law "stipulates that only with approval or acceptance of the Chinese government can foreign aircraft enter the sky above PRC territorial waters." See Gong Li, "US Plane Collision Seriously Violates International Law," *Xinhua* (April 3, 2001), as translated by Foreign Broadcast Information Service (FBIS).

10. "Aircraft Spots Possible Spratly Island Weaponry," *Defense News* (December 2–8, 1996), 2, "Woody Island Missiles," *Washington Times* (June 15, 2001).

11. A Chinese military publication, *Zhong Guo Guofang Bao* (*Chinese Defense Journal*), asserted recently that American exercises with other militaries in the region do not build confidence and are a threat to stability. See "U.S. Military Exercises with Nations in Region Worry China," *The Nation* (Thailand, August 29, 2000).

12. Rigoberto Tiglao, "Seaside Boom," *Far Eastern Economic Review* (July 8, 1999), 14.

13. Quoted in "Mahathir Says Malaysia Owns Disputed Spratly Reefs," Agence France Presse (June 27, 1999), as translated by FBIS (June 28, 1999), FBIS-EAS-1999-0627. Also see Rigoberto Tiglao, "Seaside Boom," *Far Eastern Economic Review* (July 8, 1999), 14.

14. "PRC, Malaysia Sign Joint Statement," *Xinhua* (May 31, 1999), as translated by FBIS (May 31, 1999), FBIS-CHI-1999-0531.

15. Comment by Secretary of National Defense Orlando Mercado as quoted in Greg Torode, "The Calm Before the Storm: Claimants to the Spratly Islands will Appeal to the World to help Stop China's Aggression," *South China Morning Post*, (March 14, 1999).

16. Brian Hsu, "Analysts fear PLA action in run-up to 2000 election," *Taipei Times* (September 14, 1999), 4.

17. For a discussion, see Stein Tonnesson, "Vietnam's Objective in the South China Sea: National or Regional Security," *Contemporary Southeast Asia* (April 2000).

18. Greg Torode, "The Calm Before the Storm," and "Vietnam Buys Submarines from North Korea," *Jane's Defence Weekly* (December 9, 1998), 14.

19. John H. Nocr with David Gregory, *Chokepoints: Maritime Economic Concerns in Southeast Asia* (Washington, D.C.: National Defense University, 1996), (Table 8), 26.

20. Ibid, 24.

21. Ben Barber, "Indian Navy Exercises Seen Apt to Irk Beijing: Warships Target South China Sea," *Washington Times* (May 8, 2000), 1.

22. "Indian Navy Tours East Asia," *Far Eastern Economic Review* (August 17, 2000), 10; and "Indian Warships on a Goodwill Mission to China," *The Hindu* (September 18, 2000).

23. For additional details, see Clive Schofield, "A Code of Conduct for the South China Sea?," *Jane's Intelligence Review* (November 2000), 36–40.

24. See, "China, Philippines: A Clash at Sea," *New York Times* (May 26, 1999), A6.

25. Alex L. Lacson, "War Clouds Over the Spratlys," *Business World* (November 3, 1999).

26. "Malaysian, Philippine Fighter Planes in Spratlys Standoff," Agence France-Presse (October 31, 1999), as quoted in FBIS-EAS-1999-1031 (October 31, 1999). Malaysia denied any encounter had occurred, however. Authorities in Kuala Lumpur stated that the Hawks were unarmed and flying in a different direction when they spotted the Philippine aircraft. See "No Standoff with Philippines in Spratlys: Malaysia," *Kyodo News International* (November 8, 1999); and Trevor Hollingsbee, "Spratlys Rivalry as Philippines Faces Malaysia," *Jane's Intelligence Review* (December 1999).

27. See, “Special Dispatch: Building of China's First Aircraft Carrier to Start End of This Year at Cost of 4.8 Billion Yuan,” Hong Kong *Ming Pao* (January 12, 2000), A15, as translated by FBIS (January 12, 2000); For an in-depth analysis of Chinese naval capacities and strategies, see Bernard D. Cole, *The Great Wall at Sea: China's Nava Enters the Twenty-First Century* (Annapolis, MD: Naval Institute Press, 2000).

28. For background, see Alexander Chieh-cheng Huang, “The Chinese Navy's Offshore Active Defense Strategy: Conceptualization and Implications,” *U.S. Naval War College Review* (Summer 1994), 7–32. Also see Rear Admiral Michael McDevitt, U.S. Navy (Retired), “Where is China's Navy Headed?,” *U.S. Naval Institute Proceedings* (May 2001), 58–61.

29. Dario Agnote, “China to Continue Military Buildup in Spratly Islands,” *Washington Times,* (October 22, 1999), A14.

30. Desmond Ball, “Signals Intelligence in China,” *Jane's Intelligence Review* (August 1995).

31. Dario Agnote, “China to Continue Military Buildup in Spratly Islands,” *Washington Times,* (October 21, 1999), A14.

Part II

5

Commercial Observation Satellites for Conflict Avoidance

John C. Baker

INTRODUCTION

Earth observation satellites have provided important, and sometimes vital, information on events occurring practically anywhere on the globe. During the Cold War, the United States and Soviet Union enjoyed a monopoly on high-resolution imaging satellites, which they used for monitoring each other's military forces and for verifying arms control and disarmament agreements. In parallel, the superpowers and a few other countries developed civilian remote sensing satellites, such as the U.S. Landsat and French SPOT (*Systems Probatoire d'Observation de la Terre*) satellites, which were mainly used to collect lower-resolution images of natural features and to assess the impact of human activities on the earth's surface. For the most part, these two types of Earth observation satellites—military and civilian—were focused on distinctive missions that limited their utility as an instrument for regional conflict avoidance and dispute resolution. The civilian observation satellites simply lacked the technical capabilities, such as sufficient resolution, required for effectively monitoring the objects of regional security concern (military bases or deployed forces, for example). In comparison, the military reconnaissance satellites of the major powers—the United States, Soviet Union (later Russia), and France—produced timely, high-resolution images of such objects. However, these classified images have been historically treated as state secrets that were seldom made public or even widely shared with allied governments.

The advent of high-resolution commercial and civilian observation satellites marks a new chapter in the evolution of Earth observation satellites. The new generation of non-military satellites features some major technical improvements, including the capability to collect higher-resolution imagery data compared with earlier civilian observation satellites. However, from the users' perspective, the distinction between civil and commercial observation satellite systems is eroding as both public and private imaging satellite systems provide images that can be routinely purchased by anyone, whether they are government officials, business analysts, scholars, scientists, news media organizations, or part of the general public. Both American and foreign users of imagery data will enjoy expanding access to civilian and commercial imagery because of information technology advances (more powerful personal computers and workstations, user-friendly software, and Internet-connected imagery databases, for example), which are steadily reducing the amount of specialized equipment and technical expertise required for ordering, processing, and analyzing satellite imagery data.

The growing global access to high-resolution satellite imagery has important implications for regional conflict avoidance and resolution. One of the major advantages of using commercial observation satellites is the capability to monitor the military activities of states in a way that is non-provocative and internationally accepted. Another important advantage is the ability to share these satellite images with a broad range of parties. Countries involved in regional territorial disputes, such as the Spratly Islands, can readily acquire commercial satellite imagery as a less risky way of monitoring events. In addition, they can also make the images public or even share them with their rivals without the concern of giving away military secrets because the imagery is potentially accessible to all.

Some notable precedents already exist for using satellite imagery data to help resolve territorial disputes. One precedent was the use of commercial and civilian satellite imagery to assist diplomats in resolving the difficult territorial disagreements that were impeding the negotiations of the Dayton peace accord for Bosnia in 1995. And again in the late 1990s, commercial satellite imagery played a significant role in supporting the negotiators of a peace accord between Ecuador and Peru that ended their long-standing boundary dispute. These cases illustrate the largely unanticipated benefits of using non-military satellite imagery data, which can be shared with both sides of a territorial dispute, as a diplomatic tool for moving the negotiations toward a successful conclusion.

This chapter examines the changing nature of commercial and civilian observation satellites. It reviews the key trends associated with the globalization of satellite imaging capabilities and products. Special attention is given to assessing how the new generation of non-military satellites offers technical and operational improvements that largely overcome the earlier shortcomings of non-military imaging satellites in the contribution of regional conflict avoidance and the resolution of territorial disputes.

KEY TRENDS IN CIVILIAN AND COMMERCIAL OBSERVATION SATELLITES

Since the initial earth observation satellite was launched by the United States in 1960, there has been a steady growth in the importance of space-based remote sensing for providing important data about the changing nature of the earth's surface, as well as on its atmosphere and oceans.[1] Viewing the earth from space offers special advantages—synoptic views of very large areas, and the potential for consistent and regularly updated data sets that can be used to highlight important nature features and human activity patterns. In addition, the collection of overhead images using satellites or manned spacecraft have the added political advantage of not requiring permission from the countries being observed due to the principle of "open skies" that has been accepted in practice and codified in international agreements.

The new generation of civilian and commercial observation satellites, which is becoming operational at the start of the new millennium, reflects three key trends: 1) a rapid increase in the number of countries with civilian and commercial observation satellite programs, 2) the advent of commercial remote sensing satellites, and 3) the increasingly multinational nature of remote sensing enterprises.[2] The confluence of these trends means that a growing number of non-military imaging satellites are becoming operational and that many possess the essential capabilities necessary to support regional conflict avoidance and resolution. These capabilities not only include the improving technical performance of these observation satellites but also the fact that there is unprecedented access to the imagery data that they produce among a growing number of states and non-state actors.

Internationalization

Throughout most of the Cold War years, the superpowers alone possessed the necessary resources required for developing, producing, and operating observation satellites. Their remote sensing programs built upon the major national investments in the technical expertise, organizational structures, and industrial infrastructure of the U.S. and Soviet military and civilian space programs. The substantial financial burdens and technical challenges presented by spacecraft design, launch, and operations posed hurdles that few other countries could surmount. For nearly fifteen years after the launch of the world's first civilian satellite in July 1972, the U.S. Earth Resources

Technology Satellite (or Landsat 1), other countries had no real alternative to either relying on the superpowers as the sole source of satellite imagery data or simply going without such data.

Over time, however, a growing number of other states began developing or acquiring their own civilian observation satellites for land remote sensing purposes. They were aided by the global diffusion of technical expertise and satellite manufacturing capabilities. Some countries chose to invest the substantial resources required for developing and manufacturing their own observation satellites. As shown in Table 5.1, these national programs include France's SPOT satellite series, the Japanese Earth Resources Satellite (JERS-1), and Canada's Radarsat-1 system. In addition, the European Space Agency (ESA), which coordinates the space activities of numerous European countries, produced and placed into orbit the ERS-1 and ERS-2 imaging radar satellites. However, mastery of remote sensing satellites was not confined to the industrialized countries; both China and India pursued imaging satellite programs that reached fruition by the mid-1990s.

Table 5.1

Selected Civilian Observation Satellites

Country	*Satellite system*	*Launch date*	*Imagery type*	*Resolution*
United States	Landsat 1, 2, 3, 4 (no longer operating)	1972, 1975, 1978, 1982	panchromatic and multispectral	80 meters (m) 30 m
	Landsat 5	1985	multispectral	30 m
France	SPOT 1, 2, 3 (no longer operating)	1986, 1990, 1993	panchromatic and multispectral	10 m 20 m
Canada	Radarsat 1	1995	synthetic aperture radar (SAR)	8 to 100 m
India	IRS-1C and -1D	1995 and 1997	panchromatic and multispectral	5.8 m 23.5 and 188 m
France	SPOT 4 SPOT 5	1998 planned 2002	panchromatic and multispectral panchromatic and multispectral	10 m 20 m (and 1 km) 2.5 m and 5 m 10 m
United States	Landsat 7	1999	panchromatic and multispectral	15 m 30 m
China and Brazil joint venture	CBERS-1 CBERS-2	1999 planned 2002	panchromatic and multispectral	20 m 20 to 160 m

Japan	ALOS	planned 2004	panchromatic and multispectral and SAR	2.5 m 10 m 8 m
South Korea	Kompsat-2	planned 2004	panchromatic and multispectral	1 m 4 m
Taiwan	Rocsat-2	planned 2003	panchromatic and multispectral	2 m 8 m

Note: Panchromatic is a black and white image; multispectral is a color image that includes data from both the visible and non-visible spectrum; and SAR is radar images produced by a synthetic aperture radar.

Along with the growing number of countries that have their own civilian remote sensing satellites, a much larger number of countries have acquired the necessary expertise for using remote sensing data. The leading civilian remote sensing programs, such as Landsat, SPOT, ERS, and Radarsat, have all played a major role in providing for global access to satellite imagery data and remote sensing expertise. These national programs have been a catalyst for an expanding global network of ground receiving stations along with various multinational and bilateral training activities. Their outreach efforts have accelerated the worldwide diffusion of expertise in image processing and data analysis for a broad range of scientific projects and practical applications, including resource management and urban planning. Several countries acquired the essential ground infrastructure and expertise for making use of satellite imagery data without possessing their own remote sensing satellites. These countries include Australia, Norway, Singapore, Thailand, and the United Kingdom.

The international spread of remote sensing satellites is expected to continue over the next few years as a series of Earth Observing System (EOS) satellites, such as NASA's Terra spacecraft, are launched into orbit.[3] These satellites are mainly focused on generating imagery data that is mostly useful for assessing global change such as deforestation. In comparison, several improved national satellites, such as France's SPOT 5 and Canada's Radarsat-2 satellites, will continue to be deployed. Some of these new civilian satellites will offer imaging capabilities, including higher resolution, that are more compatible with international security applications.

Some of the new civilian remote satellites will be first-time systems for several countries and even nongovernmental organizations, such as universities. These mini-satellite systems tend to be much smaller and less expensive compared to national remote sensing satellites, such as Landsat and SPOT. These satellites are being acquired by a broad range of countries, including South Korea, Taiwan, Thailand, Chile, and Algeria, with technical assistance from foreign satellite manufacturers. The smaller remote sensing systems are mainly intended to acquire greater national experience in developing and operating imaging satellites as these countries move forward with their plans to acquire more capable imaging satellite systems. However, these small satellite programs only increase the number of imaging satellites that are available to collect overhead imagery of important developments occurring almost anywhere in the world.

Commercialization

For the first time, private firms are developing their own commercial observation satellite systems, including the necessary ground infrastructure and global networks needed to market and distribute imagery data. With the end of the Cold War, the earlier government restrictions that inhibited commercial firms from developing and operating their own remote sensing satellites have been substantially relaxed. Since the mid-1990s, several American firms have received U.S. government approval to establish their own commercial remote sensing satellite operations. In addition, some non-U.S. enterprises also have been initiated in Israel and Canada with plans to develop commercial observation satellite systems.

Several firms are developing commercial observation satellites, which will produce more timely and higher-resolution imagery data that can be sold on the international market. These firms will sell raw or enhanced imagery data to a broad range of customers—governments, businesses, and nonprofit organizations. Some of the industry customers will be value-added resellers who specialize in converting the raw imagery data into information products that fit the specific needs of other customers with less expertise in working with overhead imagery. For example, some satellite imagery data is likely to be used as an important input for developing geographic information systems (GIS), which integrate a wide variety of geospatial and other data (census or land cover data, for example) into complex databases for a geographical area.

Other firms are focusing on applications, such as precision or "smart" farming, which requires imagery of moderate resolution but more frequent coverage to monitor the health of crops and to give the farmer a more precise and timely assessment of key needs, including the desired changes in fertilizer or pesticide practices. These private firms are counting on satellite imagery data to become a crucial element in information products for various commercial sectors. However, some major uncertainties exist about the nature of the worldwide market demand for satellite imagery data. Nonetheless, several U.S. and non-U.S. enterprises are proceeding on the expectation that the satellite imagery data will assume an increasingly important role in a rapidly expanding global market for geospatial information products.[4] It is quite likely that substantial changes will continue to occur among the commercial remote sensing enterprises listed in Table 5.2 in the form of diverse partnerships among different firms and even the possibility of consolidations as some firms succeed and others suffer technological setbacks or market disappointments.

Table 5.2
Selected Commercial Observation Satellites

Country	*Satellite system*	*Launch date*	*Imagery type*	*Resolution*
Russia	Resurs/SPIN-2	45-day missions (sporadic)	panchromatic and multispectral	2 and 10 m (plans for 1m)
U.S. firm Space Imaging	Ikonos (Ikonos follow-on)	1999 (planned 2004)	 panchromatic and multispectral (panchromatic) (multispectral)	 0.8 m 3 to 4 m (0.5 m) NA
U.S. firm DigitalGlobe	Quickbird	2001	panchromatic and multispectral	0.6 m 2.5 m
U.S. firm Orbimage	Orbview-3	planned 2002	panchromatic and multispectral	1 & 2 m 4 m
ImageSat International (Israel-US firm, and others)	EROS A (A1) EROS B (B1-B6)	2000 2003–2005	panchromatic panchromatic and multispectral	A = 1.8m B = .8m 3.28m
Canada (private firms and government)	Radarsat 2	planned 2003	Synthetic aperture radar (SAR)	3 m, 10 m, 25 m, 50 m, 100 m

Note: Panchromatic is a black and white image; multispectral is a color image that includes data from both the visible and non-visible spectrum; and SAR is radar images produced by a synthetic aperture radar; NA means not available.

The willingness of U.S. aerospace firms to proceed with commercial observation satellites was significantly galvanized by Presidential Decision Directive 23 (PDD-23) that President Clinton approved in March 1994. It removed many uncertainties that had been inhibiting Executive branch movement toward the licensing of commercial remote sensing satellite operations, as well as set forth governmental guidelines for foreign sales of imagery data, sensitive technologies, and even complete turn-key observation satellite systems.[5] The result was a rapid increase in the number of U.S. commercial firms planning to launch and operate imaging satellites. The new commercial imaging satellite firms are mostly American firms, except for one that originates in Israel. These commercial enterprises include the following:

- *Space Imaging, Inc.*, a U.S. commercial satellite data provider, is a spin-off firm from the Lockheed Martin Corporation. Following an initial launch failure, a second Ikonos commercial remote sensing satellite was successfully placed into orbit in September 1999, marking the true start of the world's first commercial observation satellite operations. The Ikonos imaging satellites collect both 1-meter panchromatic (black and white) imagery and 4-meter multispectral imagery, and these images of the same location can be combined into a pan-sharpened image that provides higher quality data for image interpretation and analysis. Space Imaging, Inc. has plans to build and launch two advanced imaging satellites that will be capable of collecting higher-resolution (0.5 meter) panchromatic images.

- *DigitalGlobe*, formerly known as *EarthWatch,* another commercial satellite data provider, was the early leader in receiving U.S. government approval of its commercial observation satellite plans. However, this pioneering firm suffered a major setback in December 1997 with the in-orbit failure of its EarlyBird satellite, which would have provided 3-meter panchromatic imagery, as well as the launch failure of its initial Quickbird satellite in November 2000, which would have collected 1-meter panchromatic and 4-meter multispectral imagery. The company successfully launched another Quickbird satellite with high-resolution (0.6-meter panchromatic and 2.5-meter multispectral) capabilities in October 2001.
- *Orbimage*, a U.S. commercial satellite data provider, is building on the successful launch and operation of its Orbview-2 ocean monitoring satellite, known as Seastar. Orbimage is a subsidiary of Orbital Science Corp., a space launch and technology firm that produces small launchers. Orbimage plans to launch two high-resolution satellites: Orbview-3, which will collect 1-meter panchromatic and 4-meter multispectral imagery. A similar satellite, Orbview-4, which had the same sensors plus a hyperspectral sensor being funded by the U.S. Air Force, was lost in 2001 during a launch failure.
- *ImageSat International*, a multinational enterprise formerly known as West Indies Space, involves a partnership among Israeli aerospace enterprises and a U.S. imagery database firm.[6] They have ambitious plans to build and operate up to eight commercial observation satellites known as EROS, which are derived from Israel's Ofeq (Horizon) military reconnaissance satellite system. The initial EROS-A1 satellite, which was successfully launched in December 2000, is designed to collect 1.8-meter resolution panchromatic imagery. Subsequent EROS-B satellites will collect 1-meter panchromatic images and 3.28-meter multispectral images. With its envisioned large constellation of imaging satellites, ImageSat International plans to sell various countries an exclusive right to receive its satellite imagery whenever the EROS satellites are within the downlink range of their ground receiving station.

- *Terra Server,* an American firm, is a different commercial enterprise because it involves an imagery data reseller that is focused on marketing airborne and satellite imagery data to diverse customers. It is based on a public-private partnership involving Microsoft Corporation, the U.S. Geological Survey (USGS), Aerial Images (and its SPIN-2 marketing office), the Digital Equipment Corporation, and the Eastman Kodak Company. The Terra Server database uses high-resolution panchromatic imagery (at 2- and 10-meter resolution) produced by Russian imaging satellites.

In addition, there are several civilian observation satellite programs that produce high-resolution satellite imagery, which is sold to a broad range of domestic and international customers. They include the following:

- *SPOT* (*Systeme Probatoire d'Observation de la Terre*). It is France's non-military imaging satellite program that is managed by the French space agency, the *Centre National D'Etudes Spatiales* (CNES). SPOT-1 (launched in 1986) was the first in a continuing series of SPOT remote sensing satellites that carried imaging sensors capable of collecting data with resolutions of 10 meters for panchromatic and 20 meters for multispectral images. The planned SPOT 5 system is expected to achieve a 2.5 meter resolution for its panchromatic images. Although the satellite acquisition program is government-funded, it is operated much like a commercial system in selling imagery data on the international marketplace through its commercial marketing arm known as SPOT Image.[7]
- *Russian imaging satellites.* Starting in the late 1980s, Russia assumed the leading role in selling high-resolution satellite images in the international marketplace. These images, with resolutions between 2 to 10 meters that were collected by former Soviet military reconnaissance satellites or more recent Russian imaging satellites, have been sold on the international market by a variety of domestic organizations and foreign affiliates. Although the Russian images were the highest resolution images available on the commercial market for several years, they also had the disadvantage of being produced as film products as a result of infrequent and limited duration missions for Russian imaging satellites. In mid-2000, Sovinformsputnik, a Russian enterprise that markets imagery, announced a new Moscow policy to begin selling 1-meter images despite the continuing financial problems in launching new Russian imaging satellites.
- *Japan's imaging satellites.* Japan has extensive experience in producing and operating earth observation satellites of its own. It gained valuable experience in remote sensing satellite operations with its 1992 launch of the Japan Earth Resources Satellite-1 (JERS-1), which

carried both optical sensors and a synthetic aperture radar (SAR). Japan is planning to launch a second Advanced Earth Observation Satellite (ADEOS) following the premature failure of its ADEOS-1. In addition, Japan's National Space Development Agency (NASDA) plans to place its Advanced Land Observing Satellite, which will feature both higher-resolution optical and radar imaging sensors, into orbit by 2004.

- *CBERS.* China has joined with Brazil to develop a series of earth observation satellites known as CBERS (China-Brazil Earth Resources Satellite). The first of these satellites became operational in October 1999 and a second launch is slated for 2002.[8] These satellites can collect panchromatic (black and white) and multispectral images at moderate resolution (20 meter) and broad area images at a lower resolution (260 meter).
- *IRS satellites.* India has placed into orbit a series of more capable land remote sensing satellites since its first satellite was launched in 1979. During the mid-1990s, India's IRS-1C and IRS-1D satellites offered the highest-resolution digital satellite imagery data that could be purchased on the commercial market at that time.[9] These satellites are capable of collecting panchromatic images at about 6-meter resolution and multispectral images at about 24-meter resolution. India reportedly has plans to launch in 2002 a Cartosat-1 satellite capable of collecting 2.5-meter resolution followed by a Cartosat-2 satellite with a 1-meter resolution capability in 2003.[10]
- *Radarsat.* Canada's satellite remote sensing program has focused on radar imaging satellites, which are better suited for collecting images through cloud cover and darkness. Radarsat-1, which was fully funded by the Canadian government and launched in 1995, supports a broad range of civilian monitoring requirements, including the high priority task of ice mapping and ship navigation. Radarsat's synthetic aperture radar operates in various modes that can produce imagery scenes with resolutions ranging from 100 meters for large-scale images to about 8 meters for smaller size images. The Canadian government has awarded a contract to a Canadian private firm (MacDonald, Dettwiler and Associates) to build and operate a Radarsat-2 satellite system, which is scheduled for launch in 2003. It would feature several technical improvements including the capability to collect imagery data at 3-meter resolution.

These commercial and civilian satellite programs account for about one-half of the 43 civilian and commercial remote sensing satellites that are projected to be built and launched during the first decade of the new millennium.[11] In addition, several countries are making their initial foray into satellite remote sensing by building their own smaller imaging satellites, usually with some outside technical assistance, or are acquiring them from foreign satellite manufacturers.

Multinationalization

A final trend encouraging expanded global access to satellite imagery arises from the multinational character of commercial remote sensing enterprises. Although most civilian observation satellites originated as national programs, they have generally had an international partnership component, at least in terms of a network of worldwide ground receiving stations and programs to train foreign users to make use of their imagery data. In comparison, the new commercial remote sensing operations are becoming even more multinational in character. Their multinational relationships run the gamut from business joint ventures, foreign equity investors, and strategic alliances to a wide range of international data distribution partnerships.

Faced with tight budgets and limited resources, both governmental organizations and commercial firms have a strong incentive to find international partners and equity investors to help shoulder the substantial capital and operating costs of launching and sustaining commercial remote sensing satellite operations. Another motive for these international collaborations is the marketing imperative to establish global data distribution networks. Several of the American commercial remote sensing companies have established partnerships with foreign firms to enhance worldwide marketing and distribution of their satellite's imagery data products and services. Having foreign partners can also increase access to local markets.

Hence, the confluence of major remote sensing trends (internationalization, commercialization, and multinationalization) portends the expanding access to civilian and commercial satellite imagery. Furthermore, the global availability of high-resolution satellite imagery data will not be limited to foreign governments and large corporations. In addition, a broad range of non-state and transnational actors (for example, nongovernmental organizations [NGOs], the news media, and even individuals) also will have access to satellite imagery data for the first time on a routine basis if so desired. However, despite their growing accessibility, have the new commercial and civilian observation satellites improved sufficiently to overcome the shortcomings of earlier nonmilitary satellites for supporting international security applications?

LIMITATIONS OF EARLIER CIVILIAN OBSERVATION SATELLITES FOR SECURITY APPLICATIONS

The earlier U.S. Landsat and French SPOT remote sensing satellites (refer to Table 5.1) generally lacked the technical and operational characteristics that are most desirable for regional security applications. The Landsat satellite series, which first became operational in 1972, was designed to acquire digital multispectral images of the Earth's surface areas on a regular basis. Its capability for collecting broad area scenes in different spectral bands is well suited for various civilian applications, such as scientific research, mapping, and the monitoring of natural resources, including agriculture and forests.[12] In comparison, France's SPOT satellite system was designed with a greater focus on the needs of the commercial markets, including urban planning and land use data as inputs for geographic information systems (GIS) databases. One major advantage of SPOT over Landsat is its flexible sensor pointing system for collecting images, which enables SPOT satellites to reduce the amount of time between revisit to the target location with much greater frequency.[13] Nonetheless, neither of these early generation civilian remote sensing systems was particularly well suited for international security applications where higher resolution and timeliness are needed.

The next generation of civilian and commercial observation satellites, which were launched in the mid-1990s, featured some important improvements even if they did not overcome all of the limitations of the earlier non-military imaging satellites. At 2-meter resolution, the Russian panchromatic imagery data provided by former Soviet military reconnaissance and civilian observation satellites was the highest resolution imagery that could be purchased on the commercial market for several years (Table 5.2). However, the usefulness of this high-resolution imagery for confidence-building applications has been diminished by its use of a film-based imaging system that relies on infrequent satellite launches and short-duration missions.[14]

In comparison, India's series of IRS satellites is better suited to support international security applications. The IRS-1C and IRS-1D satellites collect digital imagery data at medium resolutions (about 6 meters), which is significantly better than the resolution offered by the current Landsat and SPOT systems. Another important development was the 1995 launch of Canada's Radarsat-1. Unlike optical imaging satellites, Radarsat-1 produces radar images that are not obscured by cloud cover over the area of interest. In addition to its large area coverage capability at lower resolutions, Radarsat-1 can collect medium-resolution (about 8 meters at best) imagery data over somewhat smaller areas.

Thus, the earlier civilian observation satellites possessed capabilities that were well suited for a broad range of non-military applications. However, their capability for detecting and identifying objects of interest for international security and confidence-building applications was somewhat constrained, particularly given limits on the resolution and timeliness of the imagery that they produced. Nonetheless, even these non-military imaging satellites proved to be useful for supporting diplomatic efforts at regional conflict resolution in the 1990s.

IMPROVING CAPABILITIES OF COMMERCIAL OBSERVATION SATELLITES

The new commercial observation satellites feature a variety of improvements over the earlier civilian observation satellites. Some of these improvements make the recent imaging satellites much better suited for monitoring objects relevant to security agreements. The following are the key improvements:

- *Higher resolution.* The new commercial observation satellites are expected to include some systems that are capable of collecting imagery data with substantially higher spatial resolution (about 1 meter). Such resolution improvement will substantially increase the usefulness of the non-military satellite imagery for monitoring objects (for example, military vehicles, ships, and aircraft) and locations (for example, border areas, military base facilities, and so forth) that are relevant to international security activities.
- *Broader range of spectral capabilities.* The high-resolution imagery data offered by the new generation of commercial and civilian observation satellites are not limited to panchromatic (black and white) images. Most of the new satellites have sensors to collect multispectral (color) data, and some will even collect hyperspectral images, which can provide up to one hundred or more distinct spectral bands of imagery data. Furthermore, some firms plan to launch satellites with imaging radar systems that can collect relatively high-resolution radar imaging data.
- *Greater timeliness.* The new observation satellites also are expected to collect and distribute commercial satellite imagery data in a significantly more timely fashion. The acquisition time for new satellite imagery of desired locations could be reduced to a few days (or even hours in special cases) instead of the weeks that was often typical for acquiring data from existing civilian remote sensing satellite systems or various overseas ground receiving stations.
- *Growing numbers of satellites.* The number of commercial and civilian observation satellites in orbit could double or triple over the next few years based on current projections.[15] This substantial growth arises not only from numerous programs but also because many commercial firms and national governments plan to operate more than one remote sensing satellite to improve satellite coverage and reduce the revisit times.
- *Worldwide accessibility.* Both U.S. and foreign firms are interested in worldwide marketing of their satellite imagery with the minimum necessary restrictions imposed by their governments. In addition, they are creating global distribution networks for selling imagery data, as well as seeking multinational partnerships to facilitate market penetration in key regions, such as Europe and Asia Pacific.

With their improved capabilities, including higher resolution and growing timeliness, the new commercial observation satellites offer a potentially useful instrument to diplomats and defense planners for enhancing regional stability through conflict avoidance and greater transparency among rivals. But how would commercial satellite imagery be applied to mitigating regional conflicts? What precedents exist for the notion that these new non-military imaging satellites can significantly contribute to reducing regional conflicts?

APPLYING COMMERCIAL OBSERVATION SATELLITES TO MITIGATING INTERNATIONAL CONFLICTS

Compared with military reconnaissance satellites, which have historically played a crucial role in monitoring foreign military developments and arms control treaties, the imagery data produced by commercial observation satellites have two main advantages: 1) it can be widely shared among all countries because it is intended to be unclassified and commercially available; and 2) it is highly accessible to a broad range of potential users, including national governments, international organizations, the news media, and nongovernmental organizations. These distinguishing characteristics greatly increase the utility of commercial observation satellites in aiding mitigation of international conflicts by removing the barriers inhibiting the sharing of satellite imagery data among various countries, including rivals and potential enemies, as well as non-state actors that could assist with data sharing and third party mediation of conflicts.

The idea of applying commercial satellite imagery to mitigating regional conflicts has attracted the interest of policymakers and scholars in recent years. Some have argued that the "multiplication" of high-resolution commercial observation satellites tends to increase the opportunities for forewarning, crisis management, and conflict resolution of border disputes.[16] Although the availability of commercial observation satellites lagged behind earlier expectations, even moderate-resolution satellite imagery, such as Landsat and SPOT, has played a useful role in supporting diplomatic talks aimed at diminishing the chances of territorial conflicts. Its use was highlighted in the 1990s in the Dayton peace talks, and later in the negotiations to resolve the long-standing border conflict between Ecuador and Peru, which are briefly reviewed here.

The Dayton peace negotiations are a good example of the constructive role that satellite imagery can play in supporting regional conflict resolution. In late 1995, the United States hosted the Proximity Peace Talks that brought together the warring parties in the Bosnia-Herzegovina conflict. American negotiators assumed a proactive role in moving the Serbian, Bosnian, and Croatian delegations toward negotiating a final peace agreement. Given the complex ethnic nature of the conflict, boundaries and other territorial issues were a prominent feature of the negotiations.[17]

U.S. defense experts in mapping, satellite imagery, and geographic information systems (GIS) were assigned to support the Dayton negotiations. Their credibility in assessing the impact of the frequently proposed changes to the territorial boundaries prevented the negotiations from falling prey to technical disagreements. Both civilian satellite imagery, such as SPOT imagery, and sophisticated image processing technologies were used to assist American negotiators to resolve boundary disputes. The U.S. mediators made use of a *PowerScene* image processing system and related mapping technologies to give the negotiating parties a three-dimensional terrain visualization approach to solving seemingly intractable territorial disputes.[18] Although the application of these geospatial technologies was only one factor accounting for the success of the Dayton negotiations, it made an important contribution by eliminating territorial problems that otherwise could have stalemated the peace negotiations.

A more recent example involves American technical assistance to Ecuador and Peru in settling a stubborn border dispute over the precise location of their common boundary along the eastern slopes of the Andes and the headwaters of the Amazon. Their historical disagreement had led to sporadic border fighting, including hostilities in early 1995 that produced several hundred casualties among the troops on both sides and the loss of aircraft. Fears of an escalating conflict prompted the so-called "Guarantor Nations" (Argentina, Brazil, Chile, and the United States) to play a proactive role in supporting diplomatic negotiations between Peru and Ecuador.[19]

Along with facilitating the negotiations through quiet diplomacy, the United States made available experts and satellite imagery from the U.S. National Imagery and Mapping Agency (NIMA) to address the geospatial needs of the Ecuador-Peru negotiations. American mapping and imagery experts used both Landsat and Radarsat imagery data to create complete and more accurate maps of the border areas in this region, whose countries had lacked reliable maps because the disputed border areas are located in rugged, mountainous terrain that is frequently under cloud cover.[20] Once again, image-based terrain visualization software was used by American diplomats in assisting the negotiators from Ecuador and Peru to overcome some of the impasse to reaching agreement on a politically acceptable boundary. Following the signing of the October 26, 1998, peace accord in Brasilia, the NIMA experts also assisted Ecuador and Peru officials by helping them to produce a series of highly accurate maps to support their required border demarcation activities.

These earlier episodes suggest that civilian and commercial satellite imagery data, and their resulting information products, have a potentially valuable role to play in mitigating regional disputes. Although the United States played a major role in providing the imagery data in each case, there are increasingly fewer obstacles for other countries to assume a comparable role in providing non-military satellite imagery to the disputing countries in an effort to help contain and extinguish their dispute.

SOME CONTINUING CHALLENGES IN USING COMMERCIAL SATELLITE IMAGERY

Despite the encouraging examples of the Dayton talks and the Ecuador-Peru border negotiations, the new commercial observation satellites still possess some important constraints on their utility for supporting conflict avoidance and regional transparency. The most important include the challenge of receiving timely imagery data, the price of imagery data, and the challenge of acquiring sufficient expertise in imagery interpretation and analysis. Although none of these poses an insurmountable barrier to the use of satellite imagery, they do present at least near-term constraints.

Given that image resolution has greatly improved over the earlier Landsat and SPOT systems, the most significant shortcoming of new commercial observation satellites is the questionable timeliness of their imagery data. Under foreseeable circumstances, commercial satellite imagery will involve at least 24 to 48 hours from the time that the collection order is placed until when the imagery data is delivered to the user. This time could be considerably longer for earlier generation remote sensing satellites or film-return systems, such as the Russian observation satellite system, that place lower priority on the timely delivery of imagery data. Furthermore, for electro-optical imaging satellites, there is always the possibility that their capability to collect imagery of a specific location will be frustrated by cloud cover. Of course, not all applications for satellite imagery data depend on timely data, particularly when the imagery data is being used to address territorial issues involving natural features. However, many potential uses for military, arms control, and confidence-building missions place a premium on acquiring timely satellite imagery data.

The relatively high cost of acquiring and using satellite imagery data is another important challenge to making greater use of commercial observation satellites for international security applications. Although the costs of satellite imagery are generally decreasing as more sources of imagery data become available, current civilian and commercial observation satellites produce imagery data that can run from about $500 U.S. to over $4,000 or $5,000 U.S. per scene for imagery that is relevant to many international security applications.[21] This does not include additional charges for having the satellite make a new collection if the desired image does not already exist in the vendor's imagery database or for the use of specialized image processing and viewing equipment to prepare the imagery data for use by imagery analysts and application experts.

Finally, having technical expertise for processing and analyzing the images is still a necessity even though continuing improvements in user-friendly software makes the final information products easier to use by non-specialists. Nonetheless, high-resolution commercial satellite imagery data is unlikely to provide valid information unless the users have some degree of experience with imagery analysis and possess substantial knowledge of the historical context for interpreting the significance of objects revealed by the imagery data. Without adequate training and experience, there is a greater risk that inexperienced or new users of high-resolution satellite images will make inadvertent errors of image interpretation or analysis.

CONCLUSIONS

The new generation of non-military imaging satellites offer major improvements over the earliest Landsat and SPOT observation satellites, particularly in terms of their improved resolution and greater potential to collect worldwide images and to disseminate the resulting imagery data to users. At the same time, global accessibility to imagery products provided by the new commercial and civilian observation satellites is expanding because of reduced government restrictions, the growth of the commercial market, and new technologies, such as the Internet, that make imagery databases much more accessible to a broader range of potential users such as foreign governments and nongovernmental organizations. These trends are creating new opportunities for commercially available satellite imagery to be used in support of cooperative monitoring regimes aimed at enhancing regional stability and transparency.

NOTES

1. For a good background discussion on satellite remote sensing, see U.S. Office of Technology Assessment, *The Future of Remote Sensing from Space: Civilian Satellite Systems and Applications* (Washington, D.C.: U.S. Government Printing Office, July 1993), 1–13.
2. This discussion of remote sensing trends draws on George J. Tahu, John C. Baker, Kevin M. O'Connell, "Expanding Global Access to Civilian and Commercial Remote Sensing Data: Implications and Policy Issues," *Space Policy* 14 (August 1998), 179–188.
3. Michael Mecham, "Building on Past Missions, Terra Sets Stage for More," *Aviation Week and Space Technology* (October 4, 1999), 54–56.
4. C. Bryan Gabbard, Kevin M. O'Connell, George S. Park, and Peter J.E. Stan, *Emerging Markets of the Information Age: A Case Study in Remote Sensing Data and Technology* (Santa Monica, CA: The RAND

Corporation, 1996); and KPMG, *The Satellite Remote Sensing Industry: A Global Review* (Arlington, VA: KPMG Peat Marwick LLP, 1998), 15–17, 37–63.

5. John C. Baker, *Trading Away Security?: The Clinton Administration's 1994 Decision on Imaging Satellite Exports* (Washington, D.C.: Pew Case Studies in International Affairs, Institute for the Study of Diplomacy, Georgetown University, 1997).

6. Barbara Opall-Rome, "ImageSat International Plans Initial Public Offering," *Space News* (August 14, 2000), 16.

7. U.S. Office of Technology Assessment, *The Future of Remote Sensing from Space: Civilian Satellite Systems and Applications* (Washington, D.C.: U.S. Government Printing Office, July 1993), 1–13.

8. Carlos Santana, "The CBERS Satellites: New Earth Observation Tools," *ISPRS Highlights*, 5:2 (June 2000), 31–32.

9. Michael Mecham, "Cost-Conscious Indians Find Profits in Imaging Satellites," *Aviation Week and Space Technology* (August 12, 1996), 59–60; and Warren Ferster, "India Crafts Specialty Earth Imaging Programs," *Space News* (April 15–21, 1996), 3, 21.

10. "India Plans to Launch High-Resolution Satellite," *Space News* (November 29, 1999), 2.

11. One forecast is that about two new earth observation satellites will be annually launched over the 2000 to 2010 timeframe. Macro Caceres, "Focus Sharpens for Imaging Satellite Market," *Aerospace America* 38: 9 (September 2000), 16–18.

12. Ray A. Williamson, "The Landsat Legacy: Remote Sensing Policy and the Development of Commercial Remote Sensing," *Photogrammetric Engineering and Remote Sensing (PE&RS)* 63 (July 1997), 878–880.

13. France's SPOT (*Systeme Pour D'Observation de la Terre*) was designed with certain advantages over Landsat in terms of offering higher resolution (10 meter for panchromatic data and 20 meter for multispectral data) and a flexible viewing capability that permits the SPOT systems to collect images of the same location on the Earth's surface more frequently than Landsat.

14. The high-resolution Russian commercial satellite imaging operations rely on film rather than digital imagery, which dictates a short mission life for the observation satellites so that the film can be returned to Earth because it cannot be transmitted to a ground receiving station as can be done with digital imagery data. KPMG, *The Satellite Remote Sensing Industry: A Global Review*, 32.

15. See KPMG, *The Satellite Remote Sensing Industry: A Global Review.*

16. See the insights on this subject offered by Susan B. Chodakewsitz and Louis J. Levy, "Implications for Cross-Border Conflict," in Michael Krepon, Peter D. Zimmerman, Leonard S. Spector, and Mary Umberger, eds., *Commercial Observation Satellites and International Security* (New York: St. Martin's Press and Carnegie Endowment for International Peace, 1990), 93–97.

17. Richard Holbrooke, *To End A War* (New York: Random House, 1998), 231–312.

18. Eric Schmitt, "High-Tech Maps Guided Bosnia Talks," *New York Times* (November 24, 1995); and for an in-depth analysis, see Richard G. Johnson, "Supporting the Dayton Peace Talks," in John C. Baker, Kevin M. O'Connell, and Ray A. Williamson, eds., *Commercial Observation Satellites: At the Leading Edge of Global Transparency* (Santa Monica, CA: RAND and ASPRS, 2001), 297–310.

19. Anthony Faiola, "Peru-Ecuador Peace Treaty May Be Near," *Washington Post* (February 8, 1998); and see the diplomatic history provided by the former U.S. special envoy to the peace process in Luigi R. Einaudi, "The Ecuador-Peru Peace Process," in Chester A. Crocker, Fen Osler Hampson, and Pamela Aall, eds., *Herding Cats: Multiparty Mediation in a Complex World* (Washington, D.C.: United States Institute of Peace Press, 1999), 405–429.

20. John Gates and John Weikel, "Imagery and Mapping Support to the Ecuador-Peru Peace Process," in John C. Baker, Kevin M. O'Connell, and Ray A. Williamson, eds., *Commercial Observation Satellites: At the Leading Edge of Global Transparency* (Santa Monica, CA: RAND and ASPRS, 2001), 311–325.

21. It is worth noting that some new commercial remote sensing enterprises, such as the Terra Server project described earlier, intend to sell satellite and aerial imagery data to a broad range of consumers at relatively low prices for small image sizes, such as $30 U.S. for a one-square kilometer scene suitable for some real estate applications. However, this type of retail imagery marketing, which is mainly focused on dense population centers, will be less relevant to many international security problems that are located in relatively remote regions where relatively little commercial demand exists for high-resolution imagery coverage.

6

Cooperative Remote Monitoring in the South China Sea: Comparing the Utility of Aerial and Satellite Imagery

Vipin Gupta and Adam Bernstein

INTRODUCTION

The Spratly Islands in the South China Sea have long been a source of conflict and ill will within the region. Brunei, China, Malaysia, the Philippines, Taiwan, and Vietnam have argued for decades about sovereignty over the islands, islets, reefs, and waters in this region.[1] Some of these states have engaged in serious naval skirmishes in and around the South China Sea in protecting their sovereignty claims. They have mostly relied on a less direct but still provocative means of controlling the islands of interest. Their methods for demonstrating ownership of disputed locations include occupying uninhabited islands, constructing facilities above reefs, building landing strips, and placing territorial markers, as well as selling exploratory oil drilling rights to foreign energy exploration firms.

The competing countries often physically occupy or build structures at various locations in the Spratlys without regard to the competing legal claims to the island or reef in question. These countries appear to be trying in this way to establish a *de facto* presence, which they hope may become *de jure* over time.[2] Such tactics raise regional tensions when their competitors eventually discover the unannounced occupations. The Spratly Islands disputes often provoke naval incidents, diplomatic protests, and recurring arrests of fishermen who are operating in the disputed areas.

Unilateral efforts by the claimant countries to patrol or otherwise locally monitor the islands with aerial overflights are sporadic at best. They are made difficult by the sparse distribution of the islands, bad weather, and by the limited naval strength that most of the countries in the area have at their disposal.

In this context in which diplomatic efforts move slowly while unilateral occupation and development of the islands continues apace, it is worth considering new ways to monitor this conflict. Sensor technologies, if used to observe the South China Sea in a systematic and non-intrusive manner, might provide the necessary technical underpinnings for treaties or agreed upon codes of conduct that could defuse the multilateral conflict. While it is not the sole guarantor of the success of such initiatives, non-intrusive or remote sensing technologies can at least provide the technical basis for developing more cooperative monitoring arrangements focused on the disputed Spratly Islands.

One monitoring approach that has been suggested for use in the South China Sea is overhead imaging.[3] If changes could be seen remotely on islands and this information readily shared between the interested countries, the naval skirmishes and reprisals that have occurred in the past might be avoided or reduced in frequency. The states' knowledge that the region is under rigorous, multilateral remote surveillance could in itself discourage the unilateral development or takeover of islands or reefs in the future.

This chapter assesses the technical feasibility and usefulness of cooperative monitoring of the Spratly Islands. It compares the relative utility of using aerial reconnaissance and commercial satellite images in this role. Several islands were selected for analysis based on their political significance and the availability of archived overhead imagery. These images were analyzed to determine which features can be identified, and to examine the technical constraints placed on the regime by the small size and widespread distribution of the islands. The chapter concludes with an assessment of both the risks and benefits of remote sensing in the context of the sovereignty disputes in the region.

ANALYSIS OF AERIAL AND SATELLITE IMAGERY

Although a broad range of disputed locations among the Spratly Islands are interesting subjects for assessing the utility of aerial and satellite imagery for cooperative monitoring, an analysis of overhead imagery for only a selected location is presented in this study for illustrative purposes. The analysis presented in this chapter draws on a more in-depth analysis by the authors that applied aerial and satellite imagery to a broader range of disputed locations among the Spratly Islands.[4] In comparison, this chapter presents highlights of that analysis concerning the following key Spratly location where military outposts have been established:

- *Mischief Reef* (9° 55' N, 115° 32' E). China took control of the uninhabited reef in early 1995, triggering the Association of Southeast Asian Nations (ASEAN) to protest the surprise occupation. Aerial photos at the time indicated that China constructed several octagonal structures on the reef, which it claimed were shelters for protecting its fishermen from the elements. The Philippines, Taiwan, and Vietnam also have long-standing claims to the reef. In late 1998, China began major reconstruction activity at Mischief Reef that involved substantially expanding some of the earlier facilities into more permanent structures while dismantling some of the other octagonal structures. These relatively well-documented activities offer a good case for analyzing and comparing the utility of aerial and satellite imagery for monitoring how competing countries establish and upgrade their outposts at the disputed Spratly locations.

[*Editor's note:* The imagery analysis in this chapter uses one of the first high-resolution commercial satellite images that became available from Space Imaging's Ikonos satellite in late 1999. The Ikonos imagery data offers 0.8 meter resolution in panchromatic (black and white image) distinguishing objects on the Earth's surface. Subsequently, an even higher resolution commerical observation satellite, DigitalGlobe's Quickbird satellite, which offers a 0.61 meter resolution, became operational in late 2001, and commercial observation satellites capable of 0.5 meter resolution are planned.]

The available aerial and satellite images of Mischief Reef were analyzed for infrastructure, ongoing activities, and natural phenomena. Each image was searched for man-made and natural changes, as well as for evidence that could confirm or refute the accounts in published articles about specific details concerning the islands. The image analysis offered a means for assessing the technical capabilities and limitations of the aerial and satellite imaging systems that were used, and it gauged the monitoring utility of the new generation of commercial and civilian observation satellites that are starting to be deployed.

MISCHIEF REEF: MONITORING THE ORIGINAL CONSTRUCTION

Mischief Reef emerged as a major flashpoint in the South China Sea in early 1995 when China began constructing and occupying buildings on the reef. Filipino fishermen, who reportedly discovered the Chinese occupation, were detained by Chinese personnel at Mischief Reef, and subsequently notified Philippine authorities after being released.[6] Philippine naval and air force reconnaissance confirmed that the Chinese had taken control of the reef located 215 km west of undisputed Philippine territory.[7] Manila perceived the Chinese action as a direct challenge to its sovereignty because Mischief Reef is located within the 200 nautical mile Extended Economic Zone (EEZ) claimed by the Philippines.

The Philippine Department of National Defense reported that several Chinese naval ships were sighted in close proximity to the reef, including an amphibious warfare ship and a submarine support ship.[8] In addition to the images of the ship activity, Manila released aerial photos of four permanent building clusters on Mischief Reef that China had built.[9] Each cluster was constructed in an octagonal configuration on top of steel pylons because the reef is fully submerged at high tide. Figure 6.1 is an aerial photograph that shows one of these four building clusters at Mischief Reef in great detail. Acquired a few months after China took control of the reef, the photo shows the personnel at the building clusters carefully observing the observation aircraft as it flies by. Given its high resolution, the aerial image permits many features to be readily identified. For example, the photograph shows the armor shields placed along the periphery of the building clusters. The photo inset also shows a possibly armed individual behind one of the armor shields facing the observing aircraft.

It is useful to compare these detailed aerial photographs with the type of commercial satellite imagery that was available in the mid-1990s. Figure 6.2 is a medium-resolution satellite image of Mischief Reef that was acquired by

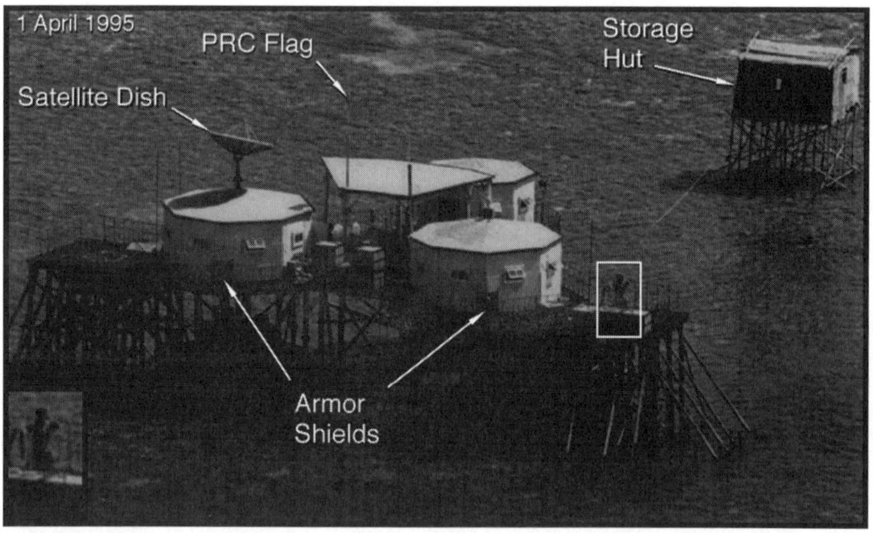

Figure 6.1

Aerial Photograph of Chinese Structure on Mischief Reef in 1995.

Source: Philippine Department of National Defense.

the Indian IRS-1C satellite in September 1997, about two and one-half years after China had erected its original facilities. This panchromatic (black and white) satellite image of Mischief Reef shows the entire landscape as well as the surrounding sea. One can readily discern the reef structure, the lagoon with a passable opening to the sea, small reef islets inside the lagoon, and the ephemeral beach along the northern periphery.

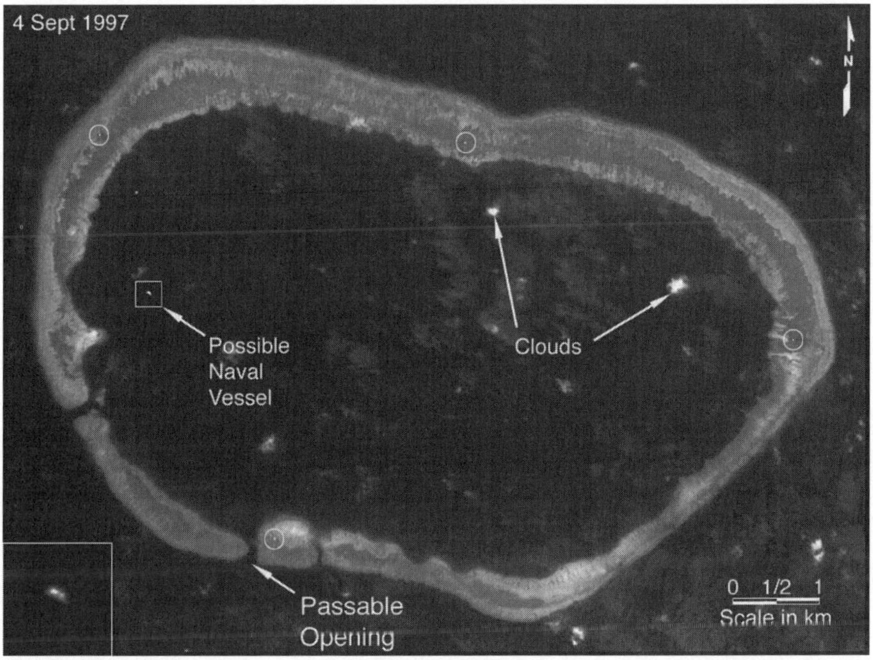

Figure 6.2

IRS-1C Satellite Image of Mischief Reef Acquired on September 4, 1997. The IRS-1C panchromatic sensor acquired the image at a six meter ground sample distance. The image was then resampled so that each pixel covers a five meter by five meter ground area. The four reported building clusters could not be identified; however, four distinct blips along the north, south, east, and west perimeters were detected. In addition, a 68 ± 5-meter long feature inside the lagoon was found that could be a naval vessel. The feature could not be conclusively identified because its shape was blurred and its shadow was not visible in the lagoon (see inset).

Source: SpaceImaging.com

However, the satellite image shows significantly less detail than the aerial photos because its 6-meter resolution (that is, ground sample distance) has a very limited ability to detect and identify smaller man-made features, such as the small Chinese octagonal building clusters. At best, the IRS-1C image shows the reef with four noticeable blips along the north, south, east, and west perimeters. These blips may be part of the four building clusters described in the media reports on the Chinese occupation. However, it is not possible to be certain about this because the individual buildings within each cluster are not visible. Inside the lagoon, there is an approximately 68-meter long feature that could possibly be a naval vessel. Unfortunately, a more definitive identification could not be made because the feature's shape is blurred and its shadow could not be discerned in the lagoon.[10]

Thus, moderate resolution, panchromatic satellite imagery provided a useful synoptic view of Mischief Reef and indicated that some type of man-made structures were on the island. However, this same imagery failed to conclusively detect all of the building clusters shown in the aerial photos. This suggests that high-resolution imagery is required to reliably detect and identify smaller scale features such as the manned outposts shown in Figure 6.1, which are representative of the smaller scale military and civilian facilities that are placed on the Spratlys by the competing countries. Such high-resolution imagery is also likely to be required to determine the nature and purpose of specific activities.

MISCHIEF REEF: MONITORING THE NEW CONSTRUCTION

The importance of high-resolution imagery was concretely demonstrated in late 1998, when the Philippine government announced renewed Chinese construction on Mischief Reef. The Philippines accused China of a substantial military buildup. China responded that the activity merely involved the repair of existing structures damaged by heavy storms. The conflicting accounts could have lingered on as an unsubstantiated factual dispute, but high-resolution photographic evidence helped to clarify the type of activity that was taking place.

Aerial photos acquired by Philippine reconnaissance aircraft show significant construction activity around the existing building clusters. Figure 6.3 shows a new, large concrete structure under construction. A five-story concrete structure was eventually completed in early 1999. High-resolution aerial photographs of the 1998–1999 construction activity at Mischief Reef clearly reveal a rapid buildup of much more permanent and fortified structures being established. Other aerial images at that time provided evidence of the substantial logistical support and protection being provided by the Chinese People's Liberation Army (PLA) navy.[11]

Figure 6.3

New Chinese Construction Activity at Mischief Reef, 1999.

Source: Philippine Department of National Defense.

Once again, it is useful to compare the aerial photographs of the construction activities occurring at Mischief Reef in 1998 with high resolution satellite imagery provided by Space Imaging's Ikonos satellite, which was launched in September 1999. The Ikonos image, which was acquired on November 26, 1999, is panchromatic with one-meter resolution (its 95 square kilometer field-of-view just covers the entire island). As expected, at one-meter resolution, satellite imagery reveals much more detail when compared with the six-meter resolution IRS-1C satellite image but far less detail than the close-up aerial photographs.

Figure 6.4 offers a close-up view of certain portions of the Ikonos image that exploits its one-meter resolution to reveal man-made structures at two of these sites, on the northern and southern sides of Mischief Reef. Individual buildings also can be clearly seen, in contrast with the earlier moderate-resolution IRS-IC satellite image (refer to Figure 6.2), in which only amorphous blips could be detected. The positions of the two built-up sites match the positions seen in the IRS-1C medium-resolution imagery. The scale of these two installations is about 100 meters, which is somewhat larger than the estimated 20-meter scale of the blips at these locations in the previous images. This apparent change in scale is consistent with the detailed records of activity shown in the aerial photographs presented in this book and by other independent reports.[12] At one-meter resolution, the Ikonos image captures the basic size and configuration of the newly expanded Chinese facilities at Mischief Reef in ways that correspond to the much more detailed view provided by the 1998 aerial photographs.

Another interesting feature of the Ikonos satellite image of Mischief Reef, which is not shown here, is that it also provides information suggesting that the two other octagonal structures erected at Mischief Reef in 1995 have probably been dismantled. In contrast to the northern and southern sites, where the new construction was detected, these features appear much less well defined.

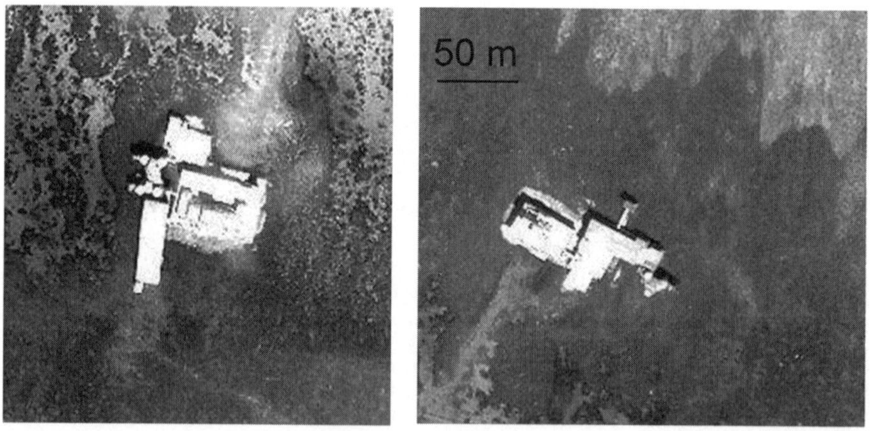

Figure 6.4

Detail of the Ikonos One-Meter Resolution Image of Mischief Reef. This Ikonos one-meter resolution satellite image of Mischief Reef was acquired on November 26, 1999. The left image shows an installation on the southern side of the reef, near the reef inlet. The right image shows a similar installation on the northern side of the reef. Individual buildings also can be easily discerned. Both installations are at or near the locations of the structures seen in the earlier medium-resolution satellite images.

Source: SpaceImaging.com

The areas seen in the Ikonos image can be compared with the blips seen in the moderate-resolution IRS-1C image in Figure 6.2. The absence of new activity at the eastern and western sites is consistent with published reports, and the presence of activity at the northern and southern sites is confirmed both by published reports[13] and by aerial photography.

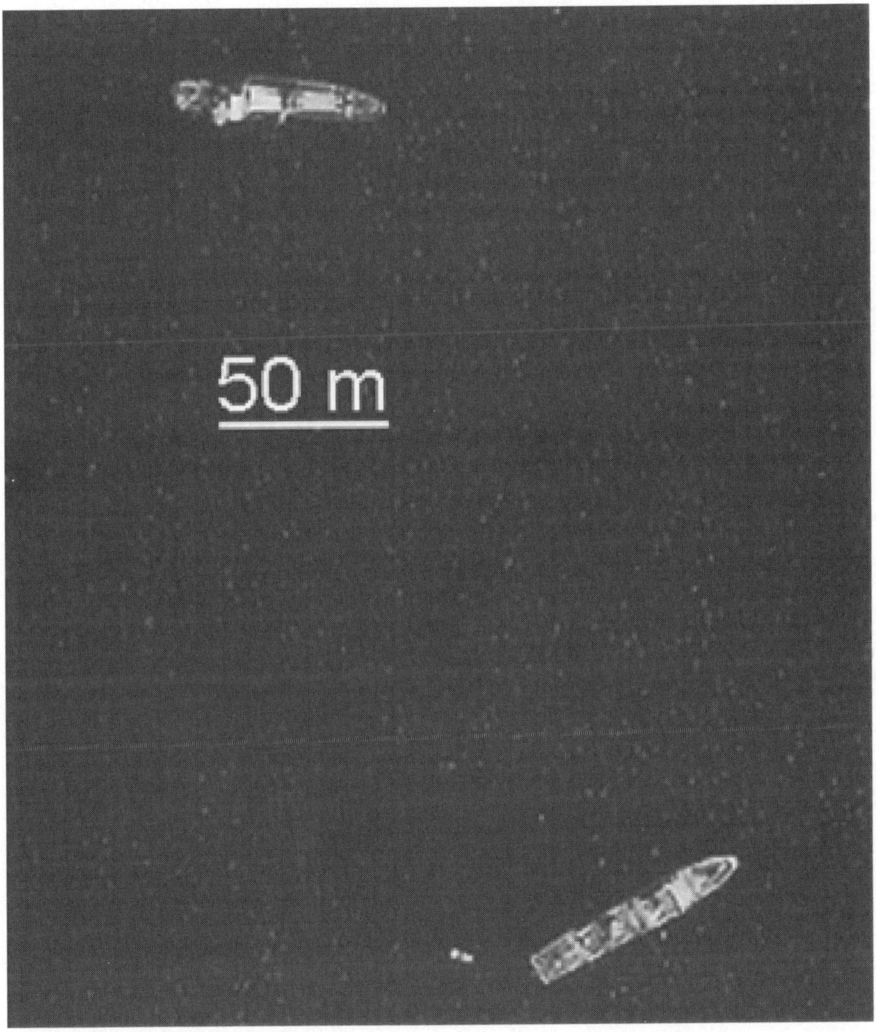

Figure 6.5

Ikonos Satellite Image of Two Ships Located in the Interior Waters at Mischief Reef.

Source: SpaceImaging.com

Finally, Figure 6.5 presents a closer look at the details of the one-meter Ikonos satellite image that reveals two smaller ships located in the interior waters of Mischief Reef. The length of these vessels, which could be Chinese supply ships, is estimated to be about 64 to 68 meters.

As the previous analysis shows, high-resolution satellite imagery is a significant addition to the set of tools available for cooperative monitoring purposes. Although the level of information that it produces does not match the information provided by a close-up aerial photograph, it is capable of removing much of the ambiguity inherent to medium-resolution satellite images and providing "good-enough" resolution for detecting key features at such small outposts in the Spratlys without many of the risks intrinsic to manned reconnaissance flights in an area of regional tension.

COMPARING AERIAL AND SATELLITE IMAGERY CAPABILITIES

Our technical analysis used a variety of aerial and satellite sensors to analyze the key disputed location among the Spratly Islands in the South China Sea. Each site revealed at least one capability or limitation associated with overhead imaging. The analysis of Mischief Reef demonstrated how high resolution aerial or satellite images were needed to detect and assess the controversial activities occurring there. Moderate-resolution imagery from satellites proved to be limited in monitoring specific activities at Mischief Reef, but it was nonetheless useful for ship detection and synoptic analysis.[14]

In addition to obtaining empirical results from the study of selected islands and reefs, we also derived results from assessing the utility of the different types of remote sensing systems that were used. Table 6.1 presents the strengths and weaknesses of four basic types of aerial and satellite imaging sensors. This assessment is based on the technical and operational utility of each sensor for monitoring the disputed locations in the South China Sea.

With the analysis in Table 6.1, we can devise a cooperative remote monitoring strategy that is optimized for the disputed Spratly Islands, as well as relevant to the broader remote sensing requirement concerning the interests of the South China Sea littoral states. The overall technical objective would be to monitor all civilian and military activity in the area without provoking an armed incident. To do this, the imaging strategy would have to rely on satellite imaging as much as possible because it is the only platform that can scan the entire South China Sea on a routine basis for ship and island activity. And, in contrast with aircraft, imaging satellites can operate from a safe altitude without the risk of being misinterpreted as a fighter-bomber.

Moderate-resolution, radar imaging satellites are the best space-based system for performing routine search missions over the South China Sea. Unaffected by adverse weather, radar imaging satellites are the only systems capable of searching for controversial activities that may deliberately use

Table 6.1

Strengths and Weaknesses of Aerial and Satellite Imaging for Monitoring the South China Sea

Remote Imaging Technology	***Strengths***	***Weaknesses***
Low Altitude (<5000 ft) Aerial Imaging	• Can acquire perspective views of ships, structures, and activity with very high spatial detail, approximately .2–.5-meter resolution (Figures 6–1 and 6–3) • Can be cued to examine suspect sites at any time of day	• Acquires images with very small fields of view and thus does not show the surrounding area • Cannot be used effectively to search for activity throughout the entire South China Sea • Cannot be used in adverse weather conditions • May be interpreted as hostile and are highly vulnerable to anti-aircraft fire
High Altitude (>5000 ft) Aerial Imaging	• Can acquire overhead images of ships, structures, and activity at high-spatial resolution, approximately .5–1-meter resolution • Can acquire perspective views at 1–5-meter resolution • Can be used to monitor a limited set of specific islands and reefs • Can be cued to examine suspect sites at any time of day	• Acquires images with small fields of view and thus does not show much of the surrounding area • Cannot be used efficiently to search for activity throughout the entire South China Sea • Cannot be used in adverse weather conditions • May be interpreted as hostile
Moderate-Resolution (5–10 m) Satellite Imaging	• Can use radar imagery to search for ships and structures throughout the South China Sea independent of weather or lighting conditions • Can use panchromatic or multispectral imagery to search for possible occupied sites and cue aerial sensors to suspect sites (see Figure 6–2)	• Cannot provide enough spatial detail to determine the nature of observed activity (see Figure 6–2) • Cannot acquire panchromatic or multispectral images of islands at night or during bad weather conditions

	• Operates at an altitude of 500–1000 km; the observed do not see the satellite that is imaging them • Are invulnerable to anti-air weaponry	• Cannot be used to image suspect sites at any time of day
High-Resolution (1–2 m) Satellite Imaging	• In addition to all the strengths of moderate-resolution satellite imaging, can show high spatial detail of ships, structures, and activity (see Figures 6–4 to 6–5)	• Cannot acquire panchromatic or multispectral images of islands at night or during bad weather conditions • Cannot be used to image suspect sites at any time • Covers a smaller area than moderate-resolution satellite images

cloud cover as a screen. Radar images from satellites such as Radarsat-1 are well suited for detecting the large ships and outposts that have been observed in the South China Sea. If the images are delivered within 8 to 72 hours of acquisition, these images could provide early warning of controversial or prohibited activity, whether civilian or military.

Any activity detected in a radar image would most likely appear as a fairly coarse blip. Consequently, high-resolution images would need to be investigated further. To facilitate the creation of a stable monitoring regime, high-resolution (one-meter or better) commercial imaging satellites would be the preferred "close look" platform because the images could be acquired from a safe and remote vantage point, transmitted to a central image archive, and shared with all claimant states and other interested parties.

If such images are unavailable in a timely fashion due to the orbital position of the satellite, the competition for tasking the satellite to collect images from other customers, or cloud coverage, then aerial imaging would be the next viable option for investigating any suspect features found in the radar satellite sweep. To minimize the risk of vulnerability and armed confrontation, aerial imaging should be done using visibly unarmed aircraft flying around the suspect site at maximum possible stand-off distances. Similarly, to maximize the degree of transparency in the South China Sea, the aerial data should be shared in a cooperative manner so that claimant states and interested parties can independently assess the observed activity. Such sharing could build on the precedent established by the Philippines' routine release of aerial images of South China Sea sites.

Our remote sensing study shows that aerial and satellite imaging can provide timely, substantive information on ships, structures, and activities in the South China Sea. With the devised optimal monitoring strategy, future studies can now consider verification provisions for specific political agreements tailored to the capabilities and limitations of aerial and satellite imaging. The essential aerial and satellite imaging platforms required for South China Sea monitoring are already operational. The number of commercial and civilian imaging satellites, and their imaging capabilities, are projected to increase over the next decade. Thus, a formal monitoring regime optimized for the South China Sea is technically feasible. Whether such a regime will be realized will ultimately depend on multilateral agreement to an interim or permanent solution to the Spratly Islands disputes.

CONCLUSIONS

This initial look at the utility of high-resolution imagery reveals the wealth of new information it can provide. The size and possible functionality of installations can be identified, and individual buildings only a few tens of meters in size can be discerned. In addition, individual ships also can be detected and in some cases identified, in contrast with the medium-resolution case, in which classes of ships could be ruled out, but conclusive identification as a rule could not be made.

In a cooperative monitoring context, high-resolution imagery can be used to demonstrate with high confidence that significant unilateral development is or is not taking place. In the particular case of monitoring the Spratly Islands, the improvement in detail makes the images both easier to analyze and easier to present and explain to adjudicating bodies and to the states party to the monitoring regime.

NOTES

1. See Chapters 2 to 4 in Part I of this book for additional background on the legal and political-military aspects of the Spratly Islands disputes.
2. Scott Snyder, *The South China Sea Dispute: Prospects for Preventive Diplomacy: A Special Report of the United States Institute for Peace* (Washington D.C.: United States Institute of Peace, August 1996), 8.
3. Scott Snyder, *The South China Sea Dispute, Prospects for Preventive Diplomacy: A Special Report of the United States Institute for Peace* (Washington D.C.: United States Institute of Peace, August 1996), 17. William Dobson and M. Taylor Fravel, "Red Herring Hegemon: China in the South China Sea," *Current History* (September 1997), 258–263.

4. The authors' more in-depth assessment, which includes additional imagery analysis of the types of Chinese naval ships observed in the area of Mischief Reef, can be found in Vipin Gupta and Adam Bernstein, "Keeping an Eye on the Islands: Cooperative Remote Monitoring in the South China Sea," in John C. Baker, Kevin M. O'Connell, and Ray A. Williamson, eds., *Commercial Observation Satellites: At the Leading Edge of Global Transparency* (Santa Monica, CA: RAND, and the American Society for Photogrammetry and Remote Sensing, 2001), 327–360.
5. Chien Chung, "Economic Development of the Islets in the South China Sea," paper delivered at the South China Sea Conference held at the American Enterprise Institute, Washington, D.C. (September 7–9, 1994), 15.
6. William Branigin, "China Takes Over Philippine-Claimed Area of Disputed Island Group," *The Washington Post* (February 11, 1995), A18.
7. "Philippine Navy Sends Ship to Disputed Shoal," *Reuters World Service* (February 2, 1995).
8. For a detailed imagery analysis of the types of Chinese naval ships that have been observed operating at Mischief Reef, see Vipin Gupta and Adam Bernstein, "Keeping an Eye on the Islands: Cooperative Remote Monitoring in the South China Sea," in John C. Baker, Kevin M. O'Connell, and Ray A. Williamson, eds., *Commercial Observation Satellites: At the Leading Edge of Global Transparency* (Santa Monica, CA: RAND, and the American Society for Photogrammetry and Remote Sensing, 2001), 334–340.
9. William Branigin, "China Takes Over Philippine-Claimed Area of Disputed Island Group," *The Washington Post* (February 11, 1995), A18.
10. The feature's shadow could not be seen because it blended with the dark lagoon water. The cloud shadows could not be seen in the lagoon either. The high sun elevation (69°) also reduced the shadow's size.
11. The imagery analysis of the Chinese Dayun- and Yuting-class amphibious warfare and naval support ships that were observed anchored off Mischief Reef in November-December 1998 can be found in Vipin Gupta and Adam Bernstein, "Keeping an Eye on the Islands: Cooperative Remote Monitoring in the South China Sea," in John C. Baker, Kevin M. O'Connell, and Ray A. Williamson, eds., *Commercial Observation Satellites: At the Leading Edge of Global Transparency* (Santa Monica, CA: RAND, and the American Society for Photogrammetry and Remote Sensing, 2001), 334–340.

12. See http://www.heritage.org/exclusive/spratly/page5.html and subsequent pages.
13. Ibid.
14. Also see the authors' satellite imagery analysis of Thitu Island, which is occupied by the Philippines. This analysis demonstrated the value of radar imagery for detecting large-scale infrastructure independent of light conditions or cloud coverage. See Vipin Gupta and Adam Bernstein, "Keeping an Eye on the Islands: Cooperative Remote Monitoring in the South China Sea," in John C. Baker, Kevin M. O'Connell, and Ray A. Williamson, eds., *Commercial Observation Satellites: At the Leading Edge of Global Transparency* (Santa Monica, CA: RAND, and the American Society for Photogrammetry and Remote Sensing, 2001), 346–348.

* This chapter originally appeared in *Commercial Observation Satellites: At the Leading Edge of Global Transparency,* edited by John C. Baker, Kevin M. O'Connell, and Ray A. Williamson, as "Keeping an Eye on the Islands: Cooperative Monitoring in the South China Sea," Chapter 16, RAND/MR-1229, Santa Monica, CA: RAND, and Bethesda, MD: American Society for Photogrammetry and Remote Sensing, 2001.

** Vipin Gupta and Adam Bernstein would like to thank John Baker, Larry Brandt, George Harris, Frank Pabian, Mark Valencia, and David Wiencek for their assistance and comments on previous drafts of the analysis presented in this chapter. They also would like to thank Tom Budge, John Olsen, and Arian Pregenzer for their guidance and support during the course of this research project. This research was supported in part by the Sandia National Laboratories. The analysis, views, and opinions expressed herein are those of the authors alone.

7

Remote Sensing Capabilities and Experience Available to the South China Sea Littoral States

Bradford L. Thomas and John C. Baker

INTRODUCTION

A key question underlying any cooperative monitoring regime for the South China Sea is whether the countries in the region have acquired an adequate level of remote sensing capability and expertise through previous experience with aerial or satellite imagery data. Governments possessing substantial remote sensing capabilities and experience are likely to be more confident in relying on the new generation of commercial and civilian observation satellites to monitor developments occurring in the Spratly Islands, particularly if they possess only limited capabilities for aerial reconnaissance or "blue-water" naval patrols. Alternatively, the littoral states of the South China Sea, with relatively limited experience in working with overhead images, could be constrained from participating in a remote sensing regime, or could be dependent on assistance from other parties.

This chapter assesses the remote sensing capabilities and experience of the littoral states of the South China Sea. It provides a survey of their civilian, commercial, and academic uses of satellite and aerial remote sensing.[1] The information provided in the chapter indicates that both the claimant and non-claimant countries concerned with the Spratly Islands have significant infrastructure for working with satellite imagery even though the regional distribution of expertise and experience is somewhat uneven. Hence, the chapter also explores the additional sources of satellite imagery expertise available to these countries that could be provided by other countries. Regional organizations and nongovernmental organizations (NGOs) that can serve as facilitators in supporting a cooperative monitoring regime for the South China Sea are also examined.

ELEMENTS OF A REMOTE SENSING INFRASTRUCTURE

In assessing the capability of the various countries in the region or beyond to use remote sensing imagery in monitoring developments in the Spratly Islands area, several elements or activities are relevant to having a robust national infrastructure for remote sensing. These include whether or not the countries being examined possess any of the following elements:

- Earth observation satellites
- Ground receiving stations for the reception of imagery data collected by the satellites[2]
- Remote sensing research and analysis centers
- Remote sensing educational and training programs
- Experience in advanced methods of imagery processing, interpretation, and analysis, including the use of imagery data as inputs to geographic information systems (GIS)

The order in which they are listed represents the descending levels of involvement in or mastery of remote sensing technology. Those higher on the list usually subsume most of the other listed activities.

SURVEY OF REMOTE SENSING EXPERTISE AND EXPERIENCE

This chapter provides a survey of the remote sensing capabilities and experience of the key countries in the South China Sea region. It is organized by 1) the remote sensing activities in the countries that are claimants to the Spratly Islands, 2) the activities of other ASEAN member countries, and 3) the potential capabilities of other governments and remote sensing businesses with interests in the South China Sea region.

China

China has developed a solid foundation for remote sensing. It possesses the most extensive capabilities and experience in remote sensing applications among the Spratly Islands claimants, even if its level of expertise is not comparable to the technologically advanced countries outside the region. The Chinese Earth Resources Observation Satellite (EROS) program has been underway for some time, having successfully launched 17 Fanhui Shi Weixing (FSW)-class recoverable imaging satellites, which were domestically produced, since 1975.[3] More recently, China has reportedly launched an imaging satellite that can return the satellite images electronically rather than using the recoverable capsules that must be de-orbited to earth.

In addition, since 1988 China has been in a partnership with Brazil to develop and deploy two improved remote sensing satellites known as the China-Brazil Earth Resources Satellites (CBERS).[4] The first satellite launch, now known as *Ziyuan 1*, occurred on October 14, 1999. This satellite has sensors to acquire panchromatic imagery at 20-meter resolution and multispectral imagery at 80-meter resolution. Of the three CBERS ground-receiving stations in China, the one in Nanning in southern China offers almost full coverage of the South China Sea. A continuing series of CBERS satellites are planned, with a second satellite launch scheduled for 2002.[5]

China operates ground receiving stations in both the Beijing area and Urumqi, in western China, which are capable of receiving Landsat data. The Beijing area remote sensing ground station at Miyun began operating in 1986 and can also receive SPOT, ERS, Radarsat, and CBERS satellite imagery data.[6] In addition, it is building a mobile ground receiving station that will be capable of receiving high-resolution data from Landsat, SPOT, and CBERS.[7] However, China's existing ground stations are not in geographical locations that will permit real-time data reception for satellite images collected of the Spratly Islands.[8]

China has three major remote sensing centers: 1) IRSA, the Institute of Remote Sensing Applications of the Chinese Academy of Science; 2) NEPA, the Chinese Research Academy of Environmental Sciences; and 3) the Institute of Remote Sensing Technology and Application at Beijing University.[9] Early remote sensing efforts in China relied on airborne instruments. The IRSA has also established both an Airborne Remote Sensing Center and a Computer Applications Center for technical support.[10] The Commission for Integrated Survey of Natural Resources carries out research on ecosystems via remote sensing. Overall, there are reportedly 400 remote sensing "units" of one kind or another throughout the nation.[11]

China is involved in technical exchanges with organizations in other countries, such as the United States Geological Survey's EROS Data Center (EDC), NASA's shuttle imaging radar program, and organizations in Australia, Canada, Japan, and the United Nations.[12] A 1992 contract with Italy's Telespazio Company calls for training of photointerpreters as well as for a provision of satellite image processing hardware and software.

Despite China's satellite launch and ground reception capabilities, its domestic technology has not met user needs. Its electronic image-processing equipment is considered not yet up to the technical requirements of high-resolution imagery, and budget shortages have hindered the domestic industry in catching up.[13] The major users in earth resources monitoring and land use mapping have been dependent on Landsat and SPOT imagery, and the meteorological users have depended on National Oceanic and Atmospheric Administration (NOAA) and Japanese GMS imagery.[14] This situation is likely to continue at least for the short term. The organizations concerned with marine environment monitoring, for example, plan to use ERS-2 radar data from the ESA. Whether China ever acquires remote sensing technology comparable to that of foreign suppliers is uncertain.

Chinese experience with higher-resolution data largely involves its domestic FSW-class photographic satellite imagery, which reportedly has 10-meter resolution, and the SPOT products, which provide 10-meter resolution panchromatic imagery and 20-meter resolution multispectral. Chinese researchers have used SPOT data in assessing the nature of urban growth patterns for China's 72 largest cities.[15] The CBERS satellite series will offer additional opportunities to work with imagery with 20-meter resolution and probably better over the longer run. Otherwise, China makes extensive use of lower-resolution imaging satellites for a broad range of remote sensing applications, including the monitoring of "red tides" that are caused by pollution in China's coastal areas.[16] China is also making progress toward placing its first oceanographic satellite, the HY-1, into orbit. This satellite will provide ocean surface data to a ground station in Sanya, in Hainan Province, and it can be used to survey ocean productivity and detect red tides.[17]

In addition, the Chinese have also had experience in making use of U.S. Shuttle Imaging Radar data, which was collected during various scientific research missions in the 1980s.[18] Cooperative projects using fine resolution radar data from the ESA's ERS-1, Canada's Radarsat, and Japan's JERS-1 will enhance Chinese expertise in interpreting detailed imagery.[19] For example, use of Radarsat imagery in tracking the Yangtse River flooding and Huang River silting was reported in 1998.[20] It also is important to note that China has a long history of utilizing airborne remote sensing imagery, not only in the collection of aerial photographic imagery, but in low-flying applications of electro-optical and radar instruments as well.[21] At such altitudes, the resulting data would certainly be high resolution.

Thus, China has the most robust remote sensing program and national capabilities among the countries directly involved in the South China Sea disputes. Nonetheless, China still lags behind the leading space countries as well as some other Asia Pacific countries, such as Japan, Australia, and possibly Singapore, in handling remote sensing areas, such as advanced image processing and analysis using digital, high-resolution satellite imagery.

Taiwan[22]

Taiwan has developed an impressive remote sensing capability and expertise of its own, which gives it a strong potential for participating in a cooperative monitoring regime. In 1984, the National Central University established the Center for Space and Remote Sensing Research (CSRSR). The Center has operated a ground receiving station capable of collecting data from Landsat, SPOT, ERS-1, and ERS-2 since 1993.[23] The ground station's coverage radius includes most of the Spratly Islands.[24] Plans for future expansion of this facility include extending its capability to be able to receive Radarsat, India's IRS-1C, and possibly other commercial observation satellites, including commercial satellite imagery. Taiwan uses both aerial and satellite imagery for a broad range of civilian purposes, including disaster assessment and recovery studies in the wake of the severe earthquake that struck Taiwan in 1999.

Taiwan is pursuing its own small experimental satellites.[25] The purpose of Republic of China Satellite No. 1 (ROCSAT 1) was scientific research and technology development.[26] Its sensor package includes a low-resolution camera for imaging ocean color, which can indicate the presence of photoplankton and pollutants. In comparison, ROCSAT 2, which is scheduled for launch in 2003, is being built with assistance from Matra Marconi Space (MMS). Although it will carry an imaging sensor capable of two-meter resolution, Taiwanese officials declare it is intended for civil and not military applications.[27]

Taiwan, perhaps alone among the Spratly Islands claimants, has already acquired experience in satellite image monitoring of those islands. In 1995, its Industrial Technology Research Institute reported having made a satellite image survey of reef locations and conditions in the Spratly Islands, which included military and other facilities erected by various claimants.[28] However, there was no indication of the type of imagery used in this exercise or its spatial resolution characteristics.

Malaysia

Malaysia possesses a modest but steadily improving remote sensing expertise that would be relevant to a cooperative monitoring regime. It does not, however, have a satellite data ground receiving station, but the Malaysian Centre for Remote Sensing (MACRES) is in the process of establishing one. This center currently provides solid professional support for the country's remote sensing activities, acquires remote sensing imagery for clients from various foreign sources, provides consultancy services for imagery users, conducts contract research in techniques and applications, and provides on-the-job training for client specialists, both private and government. MACRES also coordinates the National Remote Sensing Programme. This organization includes divisions devoted to image processing and applications, spatial data analysis, software development, and hardware development.[29]

Malaysia's remote sensing research priorities are radar remote sensing applications, the integration of remote sensing imagery and GIS, and multisensor data applications. Research has concentrated more on applications of radar than electro-optical imagery, which is better suited for imaging areas that tend to be cloud-covered much of the time. Most of its radar research has involved airborne SAR and has often depended on collaboration with foreign partners. Malaysia participated in the AIRSAR Pacific Rim campaign undertaken by NASA for research in terrain analysis, land use/land cover change detection, forest density monitoring, coastal zone data collection, and the enhancement of geological formation interpretation.[30] Malaysian specialists have also employed satellite radar imagery in a project testing ERS-1 SAR data for land cover change detection.[31]

The Philippines

Although the Philippines has developed a large, knowledgeable cadre of remote sensing experts, its civilian remote sensing program has heavily depended on outside technical assistance in developing its remote sensing capability. Following years of inadequate modernization, the Philippines remote sensing program requires an infusion of updated hardware and software, as well as renewed foreign cooperation, to overcome its growing obsolescence before it will be capable of fully participating in a cooperative monitoring regime.

Its National Remote Sensing Center (NRSC) is housed within the National Mapping and Resource Information Authority (NAMRIA) and provides digital image processing and remote sensing data to other departments of NAMRIA, as well as to other government agencies and research and academic institutions.[32] NAMRIA's Remote Sensing and Resource Data Department is specifically charged with the classification of public land as well as monitoring, assessment, and the evaluation of natural resources.[33]

The Philippines has ground receiving stations for NOAA's Advanced Very High Resolution Radiometer (AVHRR, one-kilometer resolution) and Japan's Geostationary Meteorological Satellite (GMS) data (5.1-kilometers resolution) for weather tracking and analysis.[34] However, the Philippines must depend on outside providers to collect and receive higher-resolution surface remote sensing imagery.

Remote sensing experts in the Philippines have had considerable experience in applications of SPOT imagery, both 20-meter multispectral and 10-meter panchromatic. SPOT imagery has been applied to the mapping of logging activity, forest conditions, land use, and aquaculture sites; a survey of shallow reefs; vegetation loss and land cover change around the volcano Mt. Pinatubo; land use/land cover change monitoring; environmental impact assessment mapping; basic land cover mapping to identify locust breeding environments; and 1:50,000–scale map revision.[35] Some of the change assessment projects have involved use of Landsat imagery or aerial photography as well. The change detection methodology has been visual rather than digital, though the results have usually been digitized into GIS.

Radar imagery of apparently high resolution has been used in the Philippines to assess post-earthquake lineaments, earthquake damage, geologic hazard areas, and seismic zones. In 1996, NAMRIA hosted a workshop on radar remote sensing technology and, with the Department of Science and Technology, coordinated the participation of the Philippines in NASA's AIRSAR Pacific Rim Campaign. In the latter effort, specialists from local agencies selected sites for study in geology, coastal zone conditions, soil moisture, topography, and natural hazards.[36]

Vietnam

Although the full extent of Vietnam's remote sensing experience and capabilities are not well known, it is clear that Vietnam possesses at least the core knowledge and expertise required to participate in a cooperative monitoring regime focused on the South China Sea. Growing access to outside expertise substantially bolsters Vietnam's fledgling remote sensing capability. It has received technical assistance and support from United Nations organizations, the European Union countries, and several individual governments.

According to one report, Vietnam has over 15 remote sensing centers, which are distributed among various governmental organizations or institutes.[37] While the National Committee for Space Research and Application of the National Center for Science and Technology (NCST) and its Center for Spatial Applications acts as a focal point for remote sensing activities in the country, there does not appear to be any single remote sensing center. The NCST seems to come closest to such a facility with "remote sensing centers" in its various institutes.

The General Department of Land Administration has used SPOT imagery to produce topographic maps. ERS-1 and Radarsat data have been used for natural disaster monitoring. Other remote sensing experience involves updating land use maps at medium scales, GIS development for monitoring land use changes and agriculture, medium-scale forest mapping, forest change monitoring, and forest and mangrove rehabilitation. The Integrated Resources Mapping Center (in Ho Chi Minh City) and the Forest Inventory and Planning Institute (in Hanoi) worked with the Netherlands Remote Sensing Board (BCRS) and the Coastal Zone Management Centre in the Hague on a Remote Sensing for Mangrove Rehabilitation project. This project applied SPOT satellite images to a geographical information system focused on Vietnam's Ca Mau peninsula (in the Mekong Delta) to assess the extent of the deforestation of mangrove forests, which form the transitional areas between the sea and the land that provide an important buffer against land erosion.[38]

Radarsat International (RSI), the enterprise that processes and markets Canada's RADARSAT-1 imagery has also been working with Vietnam. Taking advantage of radar imagery's special properties for monitoring rice production, RSI provided the Vietnamese Ministry of Agriculture and Rural Development with 20 RADARSAT-1 images, which covered the Mekong Delta during the 1999–2000 rice growing season, along with specialized image analysis software.[39] In addition, RSI has also been working with Vietnamese officials in developing an operational monitoring system focused on the Red River Delta that uses radar imagery for disaster management information for tracking floods and storm activities, which have struck Vietnam hard in recent years.

EXTERNAL FACILITATORS

Mistrust among the Spratly Islands claimants might well hinder the development of a cooperative monitoring regime for the South China Sea. A crucial factor in establishing such a regime could be a neutral facilitator, whether a national government or a nongovernmental or commercial organization, that could provide the acquisition of imagery, and processing and analysis of its data. Given the general reluctance in the region to submit local disputes to outside parties, a local neutral neighbor that could operate a monitoring system as an honest broker might be more acceptable. On the other hand, remote sensing expertise tends to increase away from the immediate disputed area, first in some degree with some nearby ASEAN neighbors, and then to a greater degree with some other Asia-Pacific region countries. Finally, several countries that are leaders in civilian and commercial remote sensing, such as the United States and France, could serve as international facilitators.

ASEAN

Two ASEAN member states, Thailand and Singapore, are particularly well positioned to perform such a facilitating function. Neither of these countries has any claims involving the Spratly Islands. Both countries also possess relatively sophisticated remote sensing centers with ground stations equipped to receive data from most of the earth observation satellite systems currently in operation. Both ground stations have ranges that cover the Spratly Islands.[40]

Thailand. The Thailand Remote Sensing Center ground station receives Landsat, SPOT, ERS, NOAA, MOS, and JERS-1 satellite data.[41] The Center conducts research and organizes training courses, and there is also evidence of considerable remote sensing training available at various Thai universities.[42] Thailand is also home to the Asian Institute of Technology, which has been associated with regional remote sensing training and cooperation programs discussed in later text. Thailand has a long-standing interest in acquiring its own earth observation satellite, but its plans suffered a setback with the Asian financial crisis during the late 1990s.

Singapore. The Centre for Remote Imaging, Sensing, and Processing (CRISP) ground station receives SPOT, ERS, and RADARSAT data.[43] CRISP, which was established as a research center of the National University of Singapore in 1992, was mandated to focus on regional environmental problems. CRISP conducts remote sensing studies of soil erosion, rice production, and underwater wave formation. Singapore has reportedly entered an agreement with Israel to fund joint research and development of a series of reconnaissance satellites based on Israel's Ofeq imaging satellite.[44]

Given Singapore's central role in seaborne commerce and traffic in the South China Sea, much of the remote sensing applications undertaken by CRISP are focused on monitoring maritime developments. CRISP has extensively used ERS and RADARSAT-1 imagery to monitor oil slicks (intentional oil releases from ships or unintended oil spills resulting from ship collisions, and so forth) occurring in the waters off Singapore as well as more generally in the South China Sea. CRISP has created a regional ocean pollution-mapping database for the South China Sea and related waters based on a large database of ERS satellite images to detect and measure the intensity of oil slicks found in international shipping routes.[45]

Indonesia. Indonesia is another ASEAN country that is a potential regional facilitator once its domestic political difficulties have subsided. Its National Institute of Aeronautics and Space (LAPAN) has a national remote sensing center. LAPAN and other government agencies operate five ground receiving stations that between them receive data from Landsat, SPOT, JERS, and NOAA.[46] Radarsat International is discussing with LAPAN the establishment of a ground receiving station, and has plans to conduct training in Indonesia and to establish a resource center in Jakarta. One major Indonesian remote-sensing project has used RADARSAT imagery to undertake a major coastal zone-mapping effort to support the Digital Marine Resource Mapping program. This project used the all-weather and large-area coverage capabilities of radar imagery to help identify the land-water boundaries of Indonesia's extensive coastline.[47] Other remote sensing projects have used RADARSAT and SPOT imagery to survey and monitor Indonesia's numerous active volcanoes. Through the Satellite Applications Technology Transfer in Indonesia (SATTIN), the SPOT Image Corporation is working with several Indonesia agencies to help integrate all of the earth observation processes related to satellite data reception, archiving, and distribution. As part of this three-year program, SPOT Image will provide Indonesia, with a new receiving station for SPOT 4 imagery data, a data distribution system using the Internet, space image maps of key areas of eastern Indonesia and a volcanic hazards monitoring system, as well as substantial remote-sensing training.[48]

Asia-Pacific Region

Other Asia-Pacific countries could also serve as regional facilitators, or provide additional expertise and resources to support a cooperative monitoring regime for the South China Sea. The most likely are Australia, Japan, and South Korea. Not only are these countries more removed from the dispute, both politically and geographically, but they also possess advanced levels of remote sensing technology well beyond those of the most other Asia Pacific countries.

Australia. The key element in Australia's remote sensing program is the Australian Centre for Remote Sensing (ACRES). The Centre operates a ground receiving station that receives data from Landsat, SPOT, ERS, NOAA, and JERS satellites, and has plans for RADARSAT reception.[49] The Earth Observation Centre, which is well experienced in processing and interpreting satellite imagery, is supported by Australian universities offering advanced remote sensing training. Australia has been actively involved in training remote sensing experts among the countries in Southeast Asia, including the Spratly claimant countries, as well as transferring remote sensing technology to many states.[50]

An Australian consortium, including the ACRES, has looked into building a satellite called ARIES-1 (Australian Resource Information and Environment Satellites) that will carry a hyperspectral imaging system designed for collecting imagery data on geological features and vegetation areas at 30-meter resolution along with a 10-meter panchromatic spatial resolution.[51]

Japan. Arguably the most advanced country in the Asia-Pacific region in developing and applying remote sensing technology is Japan. It has created a robust earth observation satellite program with a strong focus on environmental and land remote sensing satellites. The Japanese Earth Resources Satellite (JERS-1), which operated from 1992 to 1998, carried a synthetic aperture radar (SAR) capable of 18-meter spatial resolution.[52] In 1997, Japan launched the Advanced Earth Observation Satellite (ADEOS), an environmental monitoring satellite, into orbit, but it ceased operating after providing useful data for a few months. A replacement ADEOS satellite has been built and is being readied for launch. In addition, Japan is also working on an improved remote sensing satellite known as the Advanced Land-Observing Satellite (ALOS), which is planned for launch in 2004. It will carry a remote sensor with 2.5-meter spatial resolution and stereo-imaging capability, as well as a multispectral radiometer and an imaging radar, both with 10-meter spatial resolution.[53]

A plethora of government agencies and laboratories, quasi-governmental centers, and several universities support Japanese remote sensing research, development, and training.[54] In the Asia-Pacific region, Japan is well known for research applications of remote sensing data to local problems as well as technology transfer. Japan also is actively promoting remote sensing data exchange through a program of international cooperation in remote sensing data utilization, initially with Thailand.[55]

Stimulated by growing concern over North Korea's nuclear and missile activities, Tokyo decided in early 1999 to award a contract to build a series of high-resolution Information-Gathering Satellites (IGS) that could be used to expand Japan's satellite reconnaissance capabilities.[56] This satellite program consists of four satellites—two high-resolution (one meter) optical satellites and two high-resolution (one to three meter) radar imaging satellites. Japanese officials have insisted that these new imaging satellites are "multi-purpose" systems that will be used for a broad range of activities besides military reconnaissance, including regional disaster response and possibly supporting antipiracy operations in the South China Sea.

As Japan moves into increasingly higher-resolution earth observation satellites with collateral military reconnaissance capabilities, it may be motivated even more, in the interests of regional transparency, to share imagery and technology with neighboring countries.

South Korea. The South Korean remote sensing program has only begun to place its first earth observation satellite systems into orbit. Nonetheless, it has a strong focus on practical applications that involve national research groups that work in the forestry, fisheries, mineral exploration, and other areas. The Satellite Technology Research Center (SaTReC) has the lead responsibility in developing Korean satellites, as well as for the imaging processing systems. In May 1999, South Korea's experimental earth observation satellite, Kitsat-3, was launched into orbit. This small imaging satellite has a charged-coupled device camera (CCD) that is capable of collecting images of surface locations at 15-meter resolution.[57]

In December 1999, the Korean Multipurpose Satellite (KOMPSAT-1) was launched into orbit and delivered its imagery data to a ground station in Taejon. The primary KOMPSAT-1 sensor is an electro-optical camera capable of collecting 6.6-meter resolution digital images that can be used for mapping and surface monitoring applications such as digital elevation models (DEMs), which can be used for creating three-dimensional perspectives of terrain. A second sensor is an ocean-scanning multi-spectral imager (OSMI) with a one-kilometer resolution, which can be used for fish mapping and detecting red tides. The Korea Aerospace Research Institute (KARI) is planning for a launch of the KOMSAT-2 in 2004. This satellite will carry higher-resolution electro-optical sensors capable of collecting panchromatic images at one-meter resolution, and a Multi-Spectral Camera (MSC) capable of four-meter resolution images.[58] These imaging systems will reportedly be used for natural resource surveys, mapping, disaster monitoring, and supporting the development of South Korea's GIS. South Korea has plans for a series of KOMSAT[59] systems to be placed into orbit through 2015.

International Facilitators

Several remote sensing organizations outside the region with advanced remote sensing technology and earth resources observation satellites might possibly serve in a facilitating role. These include the U.S. National Aeronautics and Space Administration (NASA), the Canadian Space Agency,

France's CNES, the European Space Agency, and the Indian Space Research Organization (ISRO) that manages the IRS series of remote sensing satellites. While there is a general aversion in the region to involving external states in local disputes, Canada and France have been particularly active in the region in marketing their imagery and transferring technology to local countries,[60] and have developed a certain amount of trust as neutral actors. Similarly, the United States has long ties with the Asia-Pacific region based on its Landsat ground station network, the AVHHR receiving stations, and remote sensing training activities with foreign countries. Furthermore, NASA has also been directly involved in the region through programs like the Pacific Rim AIRSAR campaign[61], an airborne radar mapping activity, and the South China Sea Monsoon Experiment (SCSMEX), which integrated in situ airborne and satellite data to study summer monsoon dynamics, including precipitation structures and windfall patterns. Other countries with extensive experience in satellite and airborne remote sensing applications, such as Germany, Norway, and the United Kingdom, could also do more work with various Southeast Asia countries.

Remote sensing centers in the Asia-Pacific region, so far heavily subsidized by government organizations or academic institutions, may be following the trend toward commercialization.[62] Both commercial and government-owned or subsidized satellite imagery providers (for example, DigitalGlobe, Eurimage, ImageSat International, Orbimage, Radarsat International, SPOT Image, and Space Imaging) are increasing active in the Asia-Pacific region as part of their marketing strategies. They are selling and certifying ground receiving stations, setting up local distributorships for imagery data and products, training local specialists in processing and applying their imagery, and generally supporting remote sensing programs (sometimes even in advance of imagery coming online).[63] Although these companies are unlikely to assume a direct role in supporting a cooperative monitoring regime for the South China Sea, they could certainly be bound to play an important supporting role in providing the imagery data, as well as support imagery training and analysis.

Regional Intergovernmental and Nongovernmental Organizations

The numerous organizations in the region, both intergovernmental and nongovernmental, that are actively involved in remote sensing reflect the high level of interest in remote sensing developments and applications, which most countries in the region view as necessary for their sustainable development. The work of these transnational organizations provides an important precedent of international cooperation in remote sensing affairs, and more specifically, several of these organizations could be crucial to facilitating the operation of cooperative monitoring. An ASEAN Experts Group on Remote Sensing already exists to discuss issues in earth observation,[64] and therefore, the regional security-oriented ASEAN Regional Forum (ARF) might be an even more logical candidate for institutional support to a South China Sea cooperative monitoring regime.

One of the most prominent transnational activities on remote sensing is the Regional Space Applications Programme (RESAP) for Sustainable Development of the United Nations Economic and Social Commission for Asia and the Pacific (ESCAP), which is based in Bangkok. ESCAP actively promotes regional coordination and collaboration in remote sensing and publishes *Asia Pacific Remote Sensing and GIS Journal* covering regional remote sensing developments and activities. ESCAP has organized regional remote sensing bodies: The Regional Working Group on Remote Sensing, GIS and Space-based Positioning, and the Regional Working Group on Meteorological Satellite Applications and Natural Hazards Monitoring, which might play some role in encouraging regional cooperation on monitoring the South China Sea.[65] Similarly, the Asian Association on Remote Sensing (AARS) is a nongovernmental organization that promotes professional development, mainly through its Asian Conference on Remote Sensing (ACRS). This group of experts has endorsed the idea of developing a RICESAT, a radar satellite designed for monitoring the growth of rice in tropical Asia.[66]

The United Nations Regional Cartographic Conference for Asia and the Pacific has set up the Permanent Committee on GIS Infrastructure for Asia and the Pacific, which, especially through its charge to develop institutional frameworks for geospatial data management, could play a supportive role in a regional imagery-based monitoring system. Similarly, the countries in the Asia-Pacific region receive benefits from cooperative activities and remote sensing training provided by foreign centers. Perhaps the leading center for training specialists in remote sensing and geoinformation disciplines is the International Institute for Aerospace Survey and Earth Science, commonly known as ITC, which is based in Enschede, in the Netherlands. The ITC provides graduate and postgraduate educational programs in a broad range of geoinformation subjects, including photogrammetry, remote sensing, and GIS. The ITC also has the distinction of serving over 13,000 students since 1950.[67] This total includes over 1,000 students from the six claimant countries to the Spratly Islands. This nongovernmental organization also pursues special projects with countries in the region, including a workshop with the MACRES and other Malaysian agencies to apply remote sensing and GIS techniques to landslide hazards in June 1999, as well as a cooperative project in China to develop a coal-fire monitoring and fighting system.

CONCLUSIONS

This survey of remote sensing expertise and experience of the countries that are potential participants in a cooperative monitoring regime for the South China Sea suggests the following findings. First, a keen and rapidly growing awareness of the benefits of remote sensing technology and its applications in the South China Sea region and vicinity already exists. It is encouraged by the remote sensing activities of both regional and global organizations. Second, most of the South China Sea littoral states have acquired substantial expertise with civilian applications of satellite and aerial imagery. This should provide a sound basis for their participation in the operations of any cooperative monitoring regime concerning the Spratly Islands that makes use of high-resolution satellite imagery. Third, this regional expertise in satellite remote sensing is growing as several of the Asia-Pacific countries not only expand their applications to make use of increasingly higher-resolution imagery, but as they take steps to acquire their own higher-resolution earth observation satellite systems.

China and Taiwan, with their own satellites and ground receiving stations, seem to be the claimants to the Spratly Islands with the most advanced satellite remote sensing infrastructure. However, all of the Spratly Islands claimants, with the possible exception of Brunei, seem to possess adequate experience in applying high-resolution satellite or aerial imagery to civilian environmental or development problems. Potentially even greater expertise exists among ASEAN neighbors, such as Thailand, Singapore, and Indonesia, which have ground receiving stations covering the South China Sea. These countries could serve as regional facilitators for a cooperative monitoring regime for the Spratly Islands or the larger South China Sea. Furthermore, Australia, Japan, and even South Korea are exhibiting high levels of remote sensing expertise, with the added benefits of possessing plans for civilian observation satellites that would be able to produce imagery with the resolutions needed for effectively monitoring significant man-made activities occurring among the Spratly Islands. Multinational organizations in Europe and North America have been providing high-resolution imagery of the region for some time; however, emerging commercial firms in North America as well as Russia are likely to provide high-resolution satellite imagery (one to three meters) spatial resolution, which would be quite useful for detecting and identifying significant activities at the Spratly locations.

NOTES

1. This chapter focuses on what is publicly available regarding the civilian and commercial uses of remote sensing data by these countries and organizations. No attempt is made to assess any other imagery experience based on the use of satellite or aerial imagery for military- or intelligence-gathering purposes.

2. Given that most imaging satellites regularly collect large amounts of digital imagery data, and that these satellites possess relatively limited on-board storage capabilities, a network of ground receiving stations exists around the world to receive satellite transmissions of the imagery data whenever the satellite is operating within the "footprint" or communications cone of a particular ground station. Most national remote sensing centers are associated with these ground stations and the need to process, interpret, analyze, and archive the accumulating imagery data.

3. Brian Harvey, *The Chinese Space Programme: From Conception to Future Capabilities* (New York: John Wiley and Sons, Ltd., 1998), 52, 88–91, 152; Joan Johnson-Freese, *The Chinese Space Program: A Mystery within a Maze* (Malabar, FL: Krieger Publishing Company, 1998), 97; Johnson-Freese puts the number of FSW satellites at 14, while Harvey, in his tables, catalogues each one of the 18 (one of which, in 1993, failed on re-entry) with its date of launch. However, apparently not all of the FSW satellites carried remote sensing packages. Harvey corroborates the 10-meter spatial resolution of the cameras, but notes that the Chinese publicized the spatial resolution as 19 meters in 1996 (pp. 88–89).

4. Anne Gilks, "China's Space Policy: Review and Prospects," *Space Policy* 13:3 (August 1997), 215–227; INPE—National Institute for Space Research (Brazil) Web site: www.dgi.inpe.br/html/eng/cbers.htm, "China-Brazil Earth Resources Satellite,"; Joan Johnson-Freese, *The Chinese Space Program: A Mystery within a Maze* (Malabar, FL: Krieger Publishing Company, 1998), 97. China has also successfully launched and operated several low-resolution meteorological satellites; see Brian Harvey, *The Chinese Space Programme: From Conception to Future Capabilities* (New York: John Wiley and Sons, Ltd., 1998), 80–84.

5. Carlos Santana, "The CBERS Satellites: New Earth Observation Tools," *ISPRS Highlights*, 5:2 (June 2000), 31–32.

6. Yanping Chen, "Remote Sensing Technology and Its Utilities in China and Future International Cooperations," paper presented at the 25th International Symposium, Remote Sensing and Global Environmental Change, (Graz, Austria, April 4–8, 1993), 10; ESA—European Space Agency, ERS Web site: earth1.esrin.esa.it/ERS/, link to "ERS Satellites Ground Segment"

7. Carlos Santana, "The CBERS Satellites: New Earth Observation Tools," *ISPRS Highlights* 5:2 (June 2000), 31–32.

8. William C. Draeger, Thomas M. Holm, Donald T. Lauer, and R.J. Thompson, "The Availability of Landsat Data: Past, Present, and Future," *Photogrammetric Engineering and Remote Sensing* 63:7 (July 1997), 871. Real-time data is data received simultaneously as the satellite is passing over the target, as opposed to data that is first stored onboard the satellite and later transmitted to a ground station. The later Landsat sensors work this way.

9. IRSA—Institute of Remote Sensing Applications (China), Web site: solar.rtd.utk.edu/˜china/ins/IRSA/irsa.html, 1–2; National Research Council (NRC), *China and Global Change, Opportunities for Collaboration* (Panel on Global Climate Changes Sciences in China; Committee on Scholarly Communication with the People's Republic of China; Office of International Affairs) (Washington, D.C.: National Academy Press, 1992).

10. U.S. Foreign Broadcast Information Service (FBIS), "Science and Technology; China: New Advances in Remote Sensing" (FBIS-CST-96-002, February 12, 1996), article by Xu Guanhua and Guo Huadong (IRSA), "New Progress of Remote Sensing Science," 1–12.

11. U.S. Foreign Broadcast Information Service (FBIS), "Science and Technology; China: New Advances in Remote Sensing," (FBIS-CST-96-002, February 12, 1996), article by Xu Guanhua and Guo Huadong (IRSA), "New Progress of Remote Sensing Science," 10.

12. U.S. Foreign Broadcast Information Service (FBIS), "Science and Technology; China: New Advances in Remote Sensing," (FBIS-CST-96-002, February 12, 1996), article by Xu Guanhua and Guo Huadong (IRSA), "New Progress of Remote Sensing Science," 8–10.

13. Yanping Chen, "Remote Sensing Technology and Its Utilities in China and Future International Cooperations," paper presented at the 25th International Symposium, Remote Sensing and Global Environmental Change, (Graz, Austria, April 4–8, 1993)," 14; and Anne Gilks, "China's Space Policy: Review and Prospects," *Space Policy* 13:3 (August 1997)," 219.

14. Yanping Chen, "Remote Sensing Technology and Its Utilities in China and Future International Cooperations," paper presented at the 25th International Symposium, Remote Sensing and Global Environmental Change, (Graz, Austria, April 4–8, 1993), 13–14.

15. "Beijing SPOT Image Gets Off to a Fast Start," *SPOT Magazine*, No. 31, (2000), 16–17; and Shupeng Chen, Shan Zeng, and Chuangjie Xie, "Remote Sensing and GIS for Urban Growth Analysis in China," *PE&RS: Photogrammetric Engineering and Remote Sensing*, 66: 5 (May 2000), 593–598.

16. "PRC Employs Satellite Remote Sensing To Monitor Red Tides," *Beijing Xinhua* (April 24, 2000), reproduced in U.S. Foreign Broadcast Information Service (FBIS-CHI-2000-0424).
17. "China Building Ground Stations for Its First Oceangraphic Satellite," *Beijing Xinhua* (October 26, 2000), reproduced in U.S. Foreign Broadcast Information Service (FBIS-CHI-2000-1026).
18. Huadong Guo, "Spaceborne and Airborne SAR for Target Detection and Flood Monitoring," *PE&RS: Photogrammetric Engineering and Remote Sensing*, 66: 5, (May 2000), 611–617.
19. Yanping Chen, "Remote Sensing Technology and Its Utilities in China and Future International Cooperations," paper presented at the 25th International Symposium, Remote Sensing and Global Environmental Change, (Graz, Austria, April 4–8, 1993), 14.
20. Michael Mecham, "Imagery a Vital Planning Tool," *Aviation Week and Space Technology* 7 (December 1998), 64–69.
21. U.S. Foreign Broadcast Information Service (FBIS), "Science and Technology; China: New Advances in Remote Sensing" (FBIS-CST-96-002, February 12, 1996), article by Xu Guanhua and Guo Huadong (IRSA), "New Progress of Remote Sensing Science," 1–8.
22. Taiwan is regarded officially, by the United States as well as China, as a part of China. Its treatment here as a separate political entity in the Spratly Islands context is for convenience in examining remote sensing capabilities only and is not intended in any way as a departure from that policy.
23. CSRSR—Center for Space and Remote Sensing Research (Taiwan), Web site: www.csrsr.ncu.edu.tw.
24. William C. Draeger, Thomas M. Holm, Donald T. Lauer, and R.J. Thompson, "The Availability of Landsat Data: Past, Present, and Future," *Photogrammetric Engineering and Remote Sensing*, 63:7 (July 1997), 871.
25. Warren Ferster, "Rocsat Set to Launch Taiwan's Space Program," *Space News* (February 1, 1999), 7; NSPO—National Space Program Office (Taiwan), Web site: www.nspo.gov.tw/e30/rocsat2.html.
26. Stefan Barensky, "Taiwan Launches First Satellite," *International Space Industry Report* (February 15, 1999), 9.
27. "Science Council Chief Says No Military Use for Satellite," U.S. Foreign Broadcast Information Service (FBIS-CHI-1999-0405, March 31, 1999), 13; and "MMS to Build Taiwanese Sat," *Aviation Week and Space Technology* (December 20–27 1999), 127.

28. "Group Uses Satellite Photos To Monitor Spratlys," U.S. Foreign Broadcast Information Service (FBIS-CHI-95-073, April 17, 1995).

29. MACRES—Malaysian Centre for Remote Sensing, Web site: www.macres.gov.my.

30. *STAN—Space Technology Applications Newsletter* (ESCAP, RESAP) 15:1 (July 1997), 2–3.

31. *STAN—Space Technology Applications Newsletter* (ESCAP, RESAP) 15:1 (July 1997), 3.

32. NAMRIA—National Mapping and Resource Information Authority (Philippines), *Annual Report* (1995), 1, 12; *RSN—The Philippine Remote Sensing Newsletter* (NAMRIA) I:1 (December 1990), 4–5; *RSN* III:2 (December 1992), 9.

33. NAMRIA—National Mapping and Resource Information Authority (Philippines), *Annual Report* (1995), 5.

34. *RSN—The Philippine Remote Sensing Newsletter* (NAMRIA) III:2 (December 1992), 1.

35. *RSN—The Philippine Remote Sensing Newsletter* (NAMRIA) II:1 (July 1991), 1, 5, 10; *RSN* I:1 (December 1990), 8; *RSN* III:2 (December 1992), 7, 10; NAMRIA—National Mapping and Resource Information Authority (Philippines), *Infomapper* 1 (April 1996), 8, 10.

36. *STAN—Space Technology Applications Newsletter* (ESCAP, RESAP) 13:4 (April 1996), 4.

37. *STAN—Space Technology Applications Newsletter* (ESCAP, RESAP) 15:1 (July 1997), 5.

38. Netherlands Remote Sensing Board (BCRS), "Satellite Images for Preserving Vietnamese Mangrove Forests," Bulletin No. 32 (Delft: BCRS, February 1999).

39. "Rice Crop Monitoring System Delivered to Vietnam," *EOM: Earth Observation Magazine* (May 2000), 13.

40. CRISP—Centre for Remote, Imaging, Sensing and Processing (Singapore), Web site: www.crisp.nus.sg/crisp.html.

41. NRCT—National Research Council of Thailand, Remote Sensing Division Web site: www.nrct.go.th/htmlpages/TRSC/rs_e_1.html.

42. NRCT Web site including link to "Remote Sensing/GIS of Trang;" Internet search reveals or suggests remote sensing educational programs of some kind or other at Chiangmai University,

Chulalongkorn University, Khon Kaen University, Naresuan University, Thammasat University, and the Asian Institute of Technology.

43. Patrick Seitz, "Singapore Grows as Major Earth Imagery Center," *Space News* (October28– November 3, 1996), 28.

44. "Israel, Singapore to Sign Satellite Deal," *Jane's Defence Weekly* (July 5, 2000), 2.

45. Jingxuan Lu, "Marine Oil Spill Surveillance and Mapping Using Remote Sensing in Singapore," *Integrated Coastal Zone Management* 1 (2000), 217–222.

46. LAPAN—National Institute of Aeronautics and Space (Indonesia), Web site www.lapan.go.id/menu.html, link to "Lokasi Kerja"; *STAN—Space Technology Applications Newsletter* (ESCAP, RESAP), 13:4 (April 1996), 3.

47. Kevin P. Corbley, "Canadian Firm Provides Turnkey Solution for Indonesian Coastline Mapping Project," *EOM: Earth Observation Magazine* 9: 2 (February 2000), 25–27.

48. Igor Lampin, "The SATTIN Project: Developing Earth Observation in Indonesia," *GEO: Asia Pacific* (February–March 2000), 40–41.

49. ACRES—Australian Centre for Remote Sensing, Web site: www.auslig.gov.au/acres/index.htm; Australia also has a reception station at Hobart, Tasmania, that receives ERS and SPOT data.

50. For example, NAMRIA—National Mapping and Resource Information Authority (Philippines), *Annual Report* (1995), 12; *RSN—The Philippine Remote Sensing Newsletter* (NAMRIA) I:1 (December 1990), 4–5; "CSIRO Media Release 96/105, 24 October 1996," Web site: www.csiro.au/communication/mediarel/mr96105.htm.

51. CSIRO EOC—Commonwealth Scientific and Industrial Research Organization, Earth Observation Centre, Web site: www.eoc.csiro.au/; *Hyperspectral* means a capability to divide electromagnetic spectrum data into extremely fine slices, in this case 96 separate bands. Also see Michael Mecham, "Asia-Pacific Space: Imagery a Vital Planning Tool," *Aviation Week and Space Technology* (December 7, 1998), 68.

52. Tina Cary, "A World of Possibilities: Remote Sensing Data for Your GIS," *Geo Info Systems* 4:9 (September 1994), 38–42.

53. Steven Watkins and Paul Kallender, "Remote Sensing Faces Challenges," *Space News* (September28– October 4, 1998), 14.

54. For a good discussion of Japanese activities, see Kazuto Suzuki, "Japanese Remote Sensing Policy at a Crossroads," in John C. Baker, Kevin M. O'Connell, and Ray A. Williamson, eds., *Commercial Observation Satellites: At the Leading Edge of Global Transparency* (Santa Monica, CA: RAND, 2001), 205–223.

55. Remote Sensing Technology Center of Japan (RESTEC) link to Newsletter No. 30, "International Cooperation through The Pilot Project on Satellite Remote Sensing Data Utilization in Asian Pacific Region" at http://www.restec.or.jp/eng/library/library/html.

56. Kazuto Suzuki, "Japan Remote Sensing at a Crossroads," in John C. Baker, Kevin M. O'Connell, and Ray A. Williamson, eds., *Commercial Observation Satellites: At the Leading Edge of Global Transparency* (Santa Monica, CA: RAND, 2001), 215–220.

57. "ROK Satellite Kitsat-3 Enter Circular Orbit," Daily Report: East Asia (May 26, 1999), U.S. Foreign Broadcast Information Service (FBIS-EAS-1999-0526).

58. Barabara Opall-Rome, "Israeli Firm to Sell S. Korea 1-Meter Satellite Camera," *Space News* (March 20, 2000), 1.

59. *Beijing Xinhua* (April 24, 2000), reproduced in U.S. Foreign Broadcast Information Service (FBIS-CHI-2000-0424).

60. For example, Steven Watkins "Spot Asia Strives to Surmount Climate Hurdles," *Space News* (April 13–19), 1998.

61. NASA's Airborne Synthetic Aperture Radar (AIRSAR) was employed on a mission visiting various locations in at least five Asia Pacific countries during its Pacific Rim 2000 mission during September–October 2000. This radar imaging testbed was used to work with various countries in gathering geographic and atmospheric data for coastal analysis and oceanography, forestry, geology, hydrology and archaeology. "Flying Lab Studies Pacific Rim," *Imaging Notes* (November/December 2000), 9.

62. CSIRO EOC—Commonwealth Scientific and Industrial Research Organization, Earth Observation Centre, Web site: www.eoc.csiro.au/.

63. For example, Michael Mecham, "Imagery a Vital Planning Tool" *Aviation Week and Space Technology* 7 (December 1998); "Space Imaging Forms Japanese Partnership; Satellite Imagery Collected Locally for Pacific Rim Customers" News Releases. Web site: www.spaceimage.com/newsroom/releases/1999/japan.html;

"Orbimage Licenses NTT Data Corporation to Provide Orbview High-Resolution Satellite Imagery Services in Japan." Press Releases. Web site: www.orbimage.com/news/releases/3-23-99.html.

64. Patrick Seitz, "Singapore Grows as Major Earth Imagery Center," *Space News* (October28– November 3, 1996), 28.

65. *STAN—Space Technology Applications Newsletter* (ESCAP, RESAP) 15:1 (April 1997), 8-9

66. Shunji Murai, "Remote Sensing in Asia: Strengths, Weaknesses and Role," *GEO Asia-Pacific* (February–March 2000), 30.

67. International Institute for Aerospace Survey and Earth Sciences (ITC), *Annual Report, 1999*, (Enschede, The Netherlands, 2000), Appendix 1.4.

Part III

8

A Cooperative Monitoring Regime for the South China Sea

John C. Baker and David G. Wiencek

INTRODUCTION

As the earlier chapters in this book have discussed, a new generation of commercial and civilian observation satellites is becoming operational that will enable a much wider range of countries to monitor developments occurring among the disputed Spratly Islands. Previously, only the leading military powers, such as the United States, had imaging satellites capable of producing the high-resolution imagery needed for detecting and identifying specific human activities of political and military importance. However, increasingly over time, all countries and even nonstate actors, such as nongovernmental organizations, will have access to high-resolution satellite imagery as a growing number of new commercial and civilian observation satellites make available their imagery data. Among its many potential benefits, high-resolution satellite imagery offers a less provocative way of detecting and identifying activities of political or military importance. These activities could include when a claimant country establishes a new outpost in the Spratlys or substantially upgrades an existing one, such as occurred in 1998 at Mischief Reef.

The availability of high-resolution commercial observation satellites is an important technological prerequisite for developing a cooperative monitoring regime concerning the Spratly Islands disputes. This chapter, however, assesses the broader operational issues that will be equally important in creating a cooperative monitoring regime for mitigating the risk of armed conflict over the Spratlys. Persuading the South China Sea littoral states to participate in a cooperative monitoring regime will require more than focusing on their parochial interests in the Spratlys, even though all of these countries possess a stake in avoiding a military confrontation in the region. The regime must also address the broader interests that these countries share: the need for effective remote sensing capabilities to deal with the common challenges presented by the South China Sea. These challenges include recurring natural disasters; growing environmental problems; and the persistent problem of maritime piracy, which undermines their national sovereignty and international reputations. The political viability of a cooperative monitoring regime for the Spratly Islands will be enhanced to the extent that the monitoring mission for conflict prevention within the Spratlys is embedded in a larger pattern of collaboration among countries interested in sharing satellite imagery and other information to deal with South China Sea problems of common interest.

This chapter proposes a South China Sea cooperative monitoring regime that begins by outlining the key tenets of a cooperative monitoring strategy focused on the disputed Spratly Islands. It offers an illustrative monitoring strategy for guiding the collection of satellite imagery of the Spratly Islands, as well as estimates of the amount of imagery required on an annual basis, and the analytical capabilities required for supporting the imagery analysis. The chapter also explores the potential contribution of satellite remote sensing to the broader range of problems that the littoral states of the South China Sea have in common. Finally, the chapter highlights some of the operational elements associated with the cooperative monitoring regime and explores the relative advantages of an informal process versus a more formal structure for acquiring and sharing satellite imagery of the Spratlys.

WHY CONSIDER A TRANSPARENCY REGIME?

For the purposes of this study, a cooperative monitoring regime is an arrangement whereby parties concerned with South China Sea problems share commercial and civilian satellite imagery on either a formal or informal basis. The core goals of this proposed regime are

- To mitigate the risks of territorial conflict over the Spratlys by offering a less provocative means for monitoring politically and militarily significant developments occurring among the Spratlys compared with the current aerial reconnaissance and naval ship patrols
- To establish a broader basis for cooperation among the South China Sea littoral states for dealing with other common problems

The proposed South China Sea regime would not be limited to the various countries making claims on the Spratly Islands. Instead, this cooperative monitoring regime would be bolstered by engaging other governments and even nongovernmental organizations that have a strong interest in avoiding conflicts in the South China Sea and dealing with transnational problems, such as environmental problems and natural disasters. Thus, a potentially wide range of parties could become participants in the proposed cooperative monitoring regime, including the following:

- Spratly Islands claimants (Brunei, China, Malaysia, the Philippines, Taiwan, and Vietnam)
- Other Association of Southeast Asian Nations (ASEAN) governments (Indonesia, Singapore, Thailand, and so forth)
- Concerned external governments (Australia, Canada, France, Japan, United States, and so forth)
- Multinational organizations (ASEAN Regional Forum (ARF), European Union, and the U.N. Economic and Social Commission for Asia and the Pacific, and so forth)
- Nongovernmental organizations (the South China Sea Informal Working Group, Council for Security Cooperation in the Asia Pacific (CSCAP), International Crisis Group (ICG), and so forth)

Furthermore, nongovernmental research centers, universities, and news media organizations probably would also have access to the information being shared at some point. Using commercial satellite imagery, which can be widely shared among all interested parties, these nonstate actors could assist in monitoring the disputed Spratly Islands for early signs of any politically significant developments that might lead to an armed confrontation among the various competitors for the Spratlys.

KEY ELEMENTS OF THE SOUTH CHINA SEA COOPERATIVE REGIME

The key elements of the cooperative monitoring regime include the 1) underlying norms, 2) the monitoring strategy for both the Spratly Islands and the larger South China Sea, 3) an information collection and support system, and 4) a communication arrangement linking the parties involved in the regime.

Underlying Norms

A core tenet of the cooperative monitoring regime is the belief that the security of the South China Sea littoral states and other interested countries from outside this region will be enhanced by promoting greater regional transparency concerning activities occurring among the disputed Spratly Islands. Within the larger context of international regimes, one view holds that transparency can be defined as "the availability and accessibility of information about the regime and the performance of parties under it."[1] In this sense, transparency supports three key regime functions—coordination, reassurance, and deterrence.

Applying a transparency regime to the South China Sea disputes begins with the important question of "What is the basic norm that the regime's participants are expected to observe either through their explicit agreement or tacit cooperation?" Given the fundamental territorial disagreements over the Spratly Islands, a reasonably modest norm is required to increase the chances that the territorial claimants, particularly China, will comply with the regime expectations—even if only begrudgingly. In this case, the proposed regime norm is that each claimant should accept the idea that no new construction should occur at the islands or reefs now under their control. Nor should they start constructing structures at new locations in the South China Sea, which will be perceived by others as a very provocative act.

Essentially, the underlying norm is based on a status quo or "freeze-in-place" concept that discourages the territorial claimants from building up their human presence among the disputed Spratly Islands.[2] By extension, the norm includes the idea that the regime supporters should not make any new claims among the disputed Spratly Islands nor attempt to displace another claimant's troops from the locations they now occupy. A corollary norm is that each party accepts the right of others to observe any activities taking place on the islands and reefs among the Spratly Islands, including the activities of the occupying forces.

Given the unsettled political and military situation, a commitment by the claimant countries not to establish new outposts or substantially expand existing ones is a reasonably ambitious objective for creating a cooperative monitoring regime. This proposed regime norm has the added advantage that it is consistent with the monitoring capabilities offered by high-resolution commercial and civilian observation satellites. An alternative regime norm could be to seek demilitarization of the outposts in the Spratly Islands. Even if the claimant countries agreed in principle to this very desirable objective, they would not be able to verify that these small outposts lack any military capability without fairly intrusive means of direct access to the outpost, such as unrestricted on-site inspections by foreign delegations.

Establishing the idea of a "freeze-in-place" as the regime norm does not preclude the possibility of adopting more ambitious objectives over time if the diplomatic environment substantially improves for resolving the Spratly Islands disputes. Under the best of circumstances, the claimant countries might agree to a "rollback" concept whereby they agree to reduce their number of permanently manned outposts in the Spratlys. As the imagery analysis of Mischief Reef in Chapter 6, "Cooperative Remote Monitoring in the South China Sea: Comparing the Utility of Aerial and Satellite Imagery," illustrates, satellite imagery can provide a useful public record of whether an outpost has been established or dismantled. The existing outpost either would be fully eliminated or converted to obviously non-military functions, such as centers for eco-tourism, as has been occasionally proposed. Presumably, in this improved diplomatic climate, the sovereignty disputes over the Spratlys would be less salient and there will be little reason not to permit open access to the various Spratly locations.

Monitoring Strategy for the Spratly Islands

One of the practical aspects of a cooperative monitoring regime is to develop a satellite monitoring strategy that reassures the regime participants that none of the territorial claimants is acting contrary to the norm by quietly or covertly expanding their presence in the Spratly Islands through new or expanded outposts. An effective monitoring strategy also can offer some degree of deterrence to certain types of noncompliance with the agreed norm, particularly if the undesirable action depends on accomplishing a *fait accompli* before being detected by others. Finally, to be effective, the monitoring strategy involves some important coordination questions concerning data sharing among the countries supporting the regime.

The primary objective of the monitoring strategy for this regional transparency regime is to detect and identify politically significant activities occurring among the Spratly Islands. Our postulated monitoring strategy would focus on detecting and identifying the following types of developments:

- Any new outposts or structures established by a territorial claimant anywhere among the presently unoccupied islands, reefs, and shoals of the Spratly Islands
- Any substantial expansion to the facilities at the islands or reefs that the various claimants already occupy. Politically provocative developments can include the construction of new buildings and support facilities (for example, piers or runways for aircraft resupply operations), creating armament emplacements, undertaking oil exploration and drilling activities, and other activities that can symbolize ownership.[3]

Naturally, the monitoring strategy must be tailored to the available capabilities. Military reconnaissance aircraft and naval patrols have traditionally been the main monitoring means used by the claimant countries for keeping watch on their rivals in the Spratly Islands. These military means for obtaining information are usually well-suited for collecting relatively high-resolution imagery data and other information because of their ability to make close approaches to the disputed locations. However, they also carry some risk of provoking a limited military response, such as warning shots or worse, by the forces occuping the disputed Spratly Islands locations, particularly if they view the intruding foreign military aircraft and warships as militarily threatening during a period of increased tensions.[4] And despite the greater detail that these maritime reconnaissance forces can offer, most countries are likely to be reticent about sharing information and imagery data obtained using military means for fear of revealing the strengths or limitations of their national monitoring capabilities.

On-site inspections or visits to occupied islands and reefs by foreign observers are an alternative means of assessing whether the claimants are complying with the "freeze-in-place" norm of the regime, assuming that the observers have open access to all facilities at the site. Nonetheless, the right to conduct on-site inspection will probably require negotiating a formal agreement as part of the transparency regime—a diplomatic task that will be complicated by the sovereignty questions that it raises. Even if an agreement could be negotiated, the frequency of such visits is likely to be low because of the costs and difficulties of arranging such visits and the reluctance of the host country to accept many such intrusions. On the other hand, some claimants are more likely to invite foreign observers, including journalists, to visit the Spratly locations they occupy as a demonstration that they have nothing to hide. Unfortunately, the locations of greatest concern are unlikely to be the ones where such visits will be permitted. Overflights sponsored by nongovernmental organizaitons (NGOs) and media organizations could offer another means of acquiring a closer look at developments on the disputed islands, but these activities are likely to occur infrequently given cost constraints.

Compared with the traditional means of airborne and naval reconnaissance, satellites offer certain advantages for cooperative monitoring because they do not require permission to image anywhere on the earth. By historical precedence and international law, imaging satellites are generally not considered a provocative means of monitoring sensitive political and military developments. As noted in Chapter 5, "Commercial Observation Satellites for Conflict Avoidance," the new commercial and civilian observation satellites offer imagery data with much improved resolution and timeliness over what has been available from the earlier civilian satellites. Both commercial and civilian satellite imagery data have a major advantage because such data can be shared without the reservations associated with images obtained from military satellites or reconnaissance aircraft. In addition, commercial imagery is likely to be perceived as an impartial source of data because multiple providers of high-resolution imagery exist, and these firms have no business interest in taking sides on diplomatic issues. Thus, commercial satellite imagery possesses certain advantages for supporting a cooperative monitoring regime that is focused on developments in the Spratly Islands. These nonmilitary observation satellites offer a reasonable balance among the various technical and operational characteristics that would be desirable in a cooperative monitoring regime seeking to enhance regional transparency in the South China Sea.

Assuming that the countries bordering the South China Sea can agree on a cooperative monitoring strategy, the following would be illustrative of the type of satellite imagery collection missions that would be required:

- Routine coverage of known occupied locations
- Routine coverage of unoccupied areas
- Contingency coverage of high-interest locations

The core monitoring mission of a cooperative monitoring regime is collecting satellite imagery of the locations among the Spratlys where outposts are known to exist. Commercial and civilian satellite images would be periodically collected and analyzed for any signs of significant changes among the approximately 46 locations that are estimated to be occupied continuously by the various claimant countries. This imagery could be used to detect major expansions to existing outposts, such as the construction of new buildings, piers, landing strips for aircraft, or helicopter landing pads, as well as any signs of the deployment of crew-served weapon systems, including surface-to-air missile systems or anti-aircraft guns. On the more positive side, the routine imagery coverage also could detect and confirm if an existing Spratlys outpost has been dismantled. Higher-resolution commercial satellite imagery (one-meter or higher resolution) would be required to detect and identify these human features with a substantial confidence. As noted in Table 8.1, obtaining a high-resolution satellite image every three or four months on each of the 46 occupied Spratly locations would require annually collecting 140 to 185 images, or an average of about two cloud-free satellite images every week.

Table 8.1

Postulated Monitoring Strategy for Satellite Image Collections of the Spratly Islands

Monitoring Mission	*Preferred Type of Satellite Images*	*Annual Number of Satellite Images (Lower estimate)*	*Annual Number of Satellite Images (Higher estimate)*
Routine coverage of occupied Spratly locations (46 sites)	High-resolution optical or radar imagery	138	184
Routine broad area search of unoccupied locations located among the Spratly Islands (260,000 sq. km)	Large scene optical or radar images (typically lower resolution)	30	60

Contingency coverage (additional 10% reserve capability)	Most likely high-resolution radar or optical imagery	18	24
Total annual number of satellite images collected		186	268

Source: Authors' estimate.

Along with monitoring the known locations where outposts already exist, an equally important task involves detecting and identifying new construction of facilities at other islands or reefs that could be evidence of a claimant country taking steps to establish another manned outpost. Such changes to the status quo are particularly provocative acts and are usually accompanied by a significant increase in the level of regional tensions among certain countries. In these instances, the satellite imagery could pick up suspicious activities occurring somewhere among the disputed Spratlys locations. These activities might be signs of an initial survey team at work or even the anchorage of the large support ships used to support construction activities.

Unlike the routine coverage of occupied areas, which involves collecting images for a limited number of known locations, the construction of new outposts or related suspicious activities could occur almost anywhere among the Spratlys and their surrounding waters. This is particularly true given the propensity of countries to create outposts on reef areas. Hence, a broad area search mission is required using commercial and civilian observation satellites to cover an estimated 260,000 square kilometers among the Spratly Islands and surrounding waters where the water depth is shallow enough to support the construction of outposts. This type of monitoring mission would rely on imaging satellites that are better suited to collecting large area scenes even though these images offer relatively lower resolution. For example, SPOT, Radarsat, and possibly even Landsat satellites, which offer larger image scenes of moderate resolution, would be the preferred imagery for making spot checks of large ocean areas.

Satellite images are acceptable for broad area searches as long as they provide a sufficient level of detail to permit imagery analysts to use change detection techniques that compare images of the same location taken at two different times to detect whether any potentially important differences exist. Any discovery of new signs of probable human activities at a given location could warrant a subsequent revisit of the location with a higher-resolution satellite image to confirm or dispel concerns that a new Spratly outpost is under construction. Table 8.1 assumes that this monitoring mission would seek to acquire full coverage of the Spratly Islands area about two to four times on an annual basis based on an estimate that 15 large scene satellite images are required to cover the large area shallow water areas among the Spratlys where new outposts would most likely appear.

Finally, the monitoring strategy would include a small percentage (about 10 percent of the total imagery of contingency coverage) to provide follow-up images of high-interest locations where politically provocative activities are known or suspected to be occurring. High-resolution images could provide additional information on locations where worrisome activities have been detected through the other two types of cooperative monitoring missions. Table 8.1 estimates that about 20 to 30 additional images might be collected each year as part of this contingency coverage. These *ad hoc* satellite imagery collections are important because they offer the South China Sea littoral states a less risky alternative to undertaking more provocative military aircraft overflights or close-up naval patrols during the particularly volatile early stages of a maritime incident, which could escalate into a more serious confrontation at sea involving two or more of the countries with claims on the Spratlys.

A BROADER APPROACH FOR COOPERATIVE MONITORING OF THE SOUTH CHINA SEA

Although satellite monitoring of the Spratly Islands is desirable to enhance regional transparency, a cooperative monitoring regime must be concerned with more than the territorial disputes over the Spratly Islands and their surrounding waters. A more comprehensive monitoring approach is therefore required to address the common remote sensing needs of the South China Sea littoral states. This broader approach would take advantage of the wide range of civilian and commercial observation satellites that are increasingly available to improve the ability of these countries to deal with the following types of regional concerns:

- Scientific research on the South China Sea
- Environmental monitoring and pollution assessments
- Natural disaster warning, planning, and response
- Maritime piracy

Expanding the focus of the monitoring strategy beyond the Spratly Islands will increase the incentive for some countries, including China, to participate in a South China Sea cooperative monitoring regime. Collaborative arrangements among the countries would strengthen their remote sensing capabilities by sharing their growing expertise and pooling their limited resources. A more comprehensive monitoring strategy might also lessen any political stigma associated with monitoring the activities of the Spratly

Islands claimants by making this monitoring mission only one of several remote sensing activities concerning the South China Sea that provides benefits to all of the countries in the region.

Scientific Research

A precedent for cooperative monitoring of the South China Sea exists, although it is narrowly focused on scientific research questions. The South China Sea Monsoon Experiment (SCSMEX) is notable because it transcends regional political differences and territorial disputes to include the participation of China, Taiwan, the Philippines, Malaysia, Vietnam, as well as Indonesia, Singapore, Thailand, Australia, Japan, and the United States.[5] It seeks to better understand the physical processes of the summer monsoons in the South China Sea that greatly affect both the weather patterns and potential natural disasters in Southeast Asia and southern China. This multinational experiment, which occurred from May to June 1998, involved extensive meteorological and oceanographic observations in the northern portion of the South China Sea regarding precipitation structures, wind patterns, and other factors that influence the dynamics of summer monsoons. Along with surface observations, the experiment relied on satellite monitoring data from NASA's Tropical Rainfall Measuring Mission (TRMM) and China's meteorological satellite.[6] Improving the ability of countries to forecast the nature of the summer monsoon could help them prepare for flood damage and better design measures to minimize the loss of life and property.

Environmental Monitoring and Pollution Assessments

Environmental monitoring of the South China Sea is another potential area where cooperative monitoring using satellite imagery can offer collective benefits. Despite its vital importance as a regional source of food and livelihood, the South China Sea is suffering from a range of environmental challenges, including pollution and other human threats to sustaining its marine biodiversity over the long run.[7] The coastal areas of the littoral states are experiencing unprecedented environmental stress from a combination of growing population, rapid economic development, and increased coastal pollution, including sewage and fertilizer runoff. One result is that in recent years, the coastal regions of the Asia Pacific have suffered from instances of the so-called "red tides" or harmful algae blooms that can rapidly kill large numbers of fish, thus devastating fishing stocks and aquaculture in a region, and even posing human health problems.[8]

Similarly, coral reefs, which are considered to be the maritime equivalent of the rainforests, are at risk from overfishing, growing ship traffic, and destructive fishing practices, such as the use of cyanide and dynamite by local fishermen to increase their catches despite the long-term damage that such practices have on the fragile coral and their ability to sustain marine life over time. Coral reefs also confront an adverse environmental trend in the form of rising ocean temperatures that create "hot spots"—conditions associated with the degradation or "bleaching" of coral reefs as they begin to die and lose their color.[9] Both types of problems threaten the coral reefs of Southeast Asia, most of which are found in the coastal waters of Indonesia and the Philippines.[10]

Cooperative monitoring using civilian and commercial observation satellites is relevant to these and other regional environmental problems. In fact, remote sensing is already being used to monitor these problems as they arise. For example, the Advanced Very High Resolution Radiometer (AVHRR) sensor aboard the U.S. National Oceanic and Atmospheric Administration (NOAA) satellites was used to pinpoint hot spots in the world's oceans, including Australia's Great Barrier Reef, to help scientists better understand the conditions that contribute to coral reef bleaching.[11] Similarly, the new SeaWiFS (Sea-viewing Wide Field-of-view Sensor), on an American commercial observation satellite known as SeaStar, is collecting multispectral imagery that has been used to detect and measure red tides. By comparing images taken at different times, this satellite imagery was used to assess the nature and extent of the impact of the red tides that struck the Bohai Sea region in northern China during mid-1998.[12] Another environmental application under consideration is using imaging radars, such as the Canadian Radarsat, for monitoring the shipping lanes near coastlines for any evidence of intentional oil spills from merchant ships trying to avoid paying the costs of having their tanks cleaned out while in port.[13]

As Table 9.1 in Chapter 9, "Environmental Remote Sensing of the South China Sea," illustrates, a broad range of civilian and commercial observation satellites currently exist or are planned to be launched in the near future with capabilities relevant to monitoring coastal regions and ocean areas.[14] Although most of these space-based sensors are designed to address a variety of scientific and commercial applications, they also offer some important capabilities for environmental monitoring. In addition, most of the sensors are aboard U.S. and foreign civilian observation satellites that will provide their imagery data to qualified researchers and users for little or no direct costs. Nonetheless, making effective use of satellite imagery for monitoring environmental trends or specific problems often requires having additional sources of information, including ground truth data or on-site observations, as well as information that is seldom possessed by only one country. Hence, in some cases, data derived from satellite imagery has been integrated with geographic information systems to support coastal management and the monitoring of the health of coral reefs.[15]

Natural Disaster Warning and Response

The South China Sea littoral states also can take advantage of satellite imagery data to deal with the natural disasters that regularly occur in this region. Many of these countries suffer each year from the effects of major floods, earthquakes, and typhoons that result in numerous deaths, displaced local populations, and substantial property damage. And some countries are occasionally afflicted by deadly tsunamis and large volcanic eruptions, such as the widespread local devastation caused by the 1991 eruption of Mt. Pinatubo in the Philippines.[16] These countries are particularly vulnerable to the destructive power of major natural disasters, and have a strong incentive to minimize the costs and disruptions imposed on their economic development by natural disasters.

Civilian and commercial observation satellites are particularly well suited to producing precise and timely geospatial information that can assist national and local officials in dealing with the adverse societal effects of large-scale natural disasters. Satellite imagery provides an important data source for national and multinational disaster management organizations responsible for anticipating major natural disasters, estimating their likely impact on humans, providing timely warnings to the specific populations at risk, assessing the extent of damage in the aftermath of the disaster, and supporting emergency response and recovery operations. Furthermore, a broad range of imaging satellites are likely to be relevant to this remote sensing mission, including weather satellites, civilian observation satellites, particularly radar imaging satellites, and the new higher-resolution commercial observation satellites.

Some of the capabilities of imaging satellites for monitoring are well-known, such as using optical imagery from Landsat, SPOT, and other civilian satellites to assess the extent of flood or earthquake damage.[17] The multispectral capabilities of these optical satellites can help emergency planners to assess the increased vulnerability of regions to flooding and mudslides due to the effects of deforestation and inadequate soil erosion control practices. Imaging radar systems, such as Radarsat, are particularly well-suited for monitoring floods because of the radar's ability to penetrate cloud cover.[18] And new imaging techniques, such as radar interferometry, which can be used to detect very small changes in the earth's surface, offer a promising means for monitoring and anticipating major changes in volcanic activity and even earthquakes.

Several countries, such as Japan and France, have been strong advocates of developing global information networks for collecting and distributing satellite imagery data to assist national preparations and emergency responses to natural disasters.[19] In combination, the broad range of existing and planned civilian and commercial observation satellites will offer a robust remote sensing capability for providing timely geospatial information relevant to a wide range of natural disasters that regularly occur in the South China Sea region.

Maritime Piracy

Satellite imagery data can also be used to help the national authorities of the South China Sea littoral states in their fight against maritime piracy. Losing ships or their cargoes to piracy is an age-old maritime hazard that has acquired a new face in recent years.[20] Numerous incidents of modern piracy occur each year, particularly in the East China Sea, South China Sea, and the Gulf of Thailand. The targets of these pirate attacks are usually commercial cargo ships and oil tankers.

As Chapter 10, "Contemporary Maritime Piracy in Southeast Asia," discusses in greater detail, modern day piracy can range from nuisance attacks by local pirates who simply rob the unfortunate crews of whatever ship they board to much more sophisticated criminal operations. The latter involves relatively well-armed groups with inside information and good connections that hijack ships carrying valuable cargoes, such as luxury automobiles or expensive minerals, and then set their crews adrift (or even murder them) and sell off the contents of the cargoes, usually with the complicity of local authorities. In some cases, such as southern China, local officials and even the Chinese military are suspected of supporting or at least tolerating the piracy.[21] Hence, transiting the South China Sea can be a risky business for many commercial and private vessels, and one that affects the safety and freedom of the seas for all countries.

Satellite imagery is a potentially important tool for helping maritime officials in the countries surrounding the South China Sea to combat piracy by monitoring the shipping lanes and tracking down ships that have been hijacked. Civilian imaging satellites, such as Radarsat, have sufficient resolution and all-weather capability to monitor the sea lanes with a reasonable degree of confidence as a supplement to other sources of maritime information. In addition, the high-resolution optical satellites could provide imagery evidence that hijacked ships are moored at particular locations or ports, which could assist the owners and insurance companies in securing government assistance in recovering the hijacked vessels.

Information Collection and Support System

The operational heart of any cooperative monitoring regime is the information collection and support system that the regime participants would rely on for acquiring, processing, and analyzing satellite imagery data. In the case of a cooperative monitoring regime for the South China Sea, two basic approaches are conceivable:

- An informal process whereby individual countries or (non-state actors) acquire and share imagery data at their own discretion
- A formal structure for imagery data sharing based on an explicit agreement among the participants in the transparency regime

Informal Process. The informal process operates largely on an *ad hoc* basis. Governments or non-state actors, such as NGOs, would decide on their own whether to acquire commercial satellite imagery of disputed locations among the Spratly Islands. They would also unilaterally decide whether to share their findings and raw imagery data with other interested countries. While this approach gives national governments substantial leeway in making monitoring decisions, it also imposes substantial transaction costs on them because they would carry the burden of purchasing the satellite imagery and undertaking the necessary image processing and analysis on their own. The findings could be shared in a confidential manner or more openly through images and analysis publicly released through the news media and on the Internet as part of a public diplomacy effort. Other governments and nonstate actors could assess the validity of the imagery analysis by purchasing their own copy of the commercial satellite image and then processing and analyzing the imagery data for themselves or with some assistance from outside experts.

Formal Structure. An alternative approach could result from a formal agreement among the various countries that are willing to support a transparency regime focused on the Spratly Islands. This agreement would use a more structured process to implement a monitoring strategy for the disputed Spratly Islands, as well as to assess the implications of any politically sensitive developments occurring at the islands and reefs that might be detected by analyzing satellite imagery. The necessary information collection and support system would rely on a small, multinational cadre of imagery experts operating under the auspices of the ARF. The team of specialists would make available their imagery analysis findings (along with the raw imagery data) to the ARF member states, as well as any other entities supporting the transparency regime, thus reducing the transaction costs for all parties.

We estimate that only a small staff of imagery experts and administrative support personnel would be needed to support a cooperative monitoring regime for the South China Sea. The composition and key responsibilities of this cadre of imagery experts could be the following:

- Postulated Imagery Analysis Staff for handling about 185 to 270 images per year: 4 professional imagery experts and 2 support personnel:
    - One Staff Director (interfaces with policymakers, and focuses on budget, and personnel issues)
    - One Assistant Director/Chief Imagery Expert (day-to-day management of imagery collection and analysis activities)
    - Two Staff Imagery Experts (experts in imagery processing, interpretation, and analysis)
    - Two Support Staff Personnel (administrative and archiving tasks)

  - Consultants as necessary on imagery analysis, geography of the Spratly Islands,and so forth

- Postulated key responsibilities of the Imagery Analysis Staff are as follows:
  - Acquiring civilian and commercial imagery (tasking, archives search, and ordering)
  - Performing basic imagery processing (as required)
  - Performing imagery analysis and presentations to sponsoring agency
  - Producing imagery analysis reports
  - Disseminating imagery analysis reports and imagery data to transparency regime participants (government and nonstate parties)
  - Consulting with technical experts in supporting governments and NGOs

We estimate that about $50,000 U.S. of imagery-handling equipment would be required to support the postulated Imagery Analysis Staff, along with about $300,000 to $500,000 U.S. in annual personnel costs to support the small staff.[22] The added costs of acquiring 185 to 270 commercial and civilian satellite images would increase the direct costs of the satellite monitoring operations to about $750,000 to $1 million U.S. annually. These costs could be reduced if this staff was able to draw on the imagery data receiving and processing capabilities of an impartial country for a nominal charge. Australia, Singapore, and Thailand all possess substantial remote sensing expertise and infrastructure, but none are territorial claimants to the Spratly Islands disputes. Similarly, civilian and commercial satellite data providers might be willing to provide their imagery data at significantly discounted rates for diplomatic or even commercial reasons. Otherwise, the collection and analysis of satellite imagery might be financed by the collective contributions of the countries participating in the Spratly Islands transparency regime.

Of course, these approaches are not mutually exclusive because a transparency regime for the Spratly Islands might begin as an informal process and eventually evolve into a more formal structure on the basis of practice, and perhaps eventually, a negotiated agreement. Each approach has its own advantages and shortcomings (see Table 8.2). For example, the informal process can begin without reaching agreement with all parties in advance while the formal structure offers greater potential for formulating a more coherent monitoring strategy and has the great advantage of spreading the transaction costs among a large number of participants.

Communication Network. A transparency regime for the Spratly Islands requires not only the information provided by satellite imagery and other monitoring sources, but also a communications network for linking its par-

Table 8.2

Alternative Approaches for Satellite Monitoring of the Spratly Islands

	Informal Process	***Formal Structure***
Key features	• Norms based on tacit understandings • Imagery collection and analysis performed by individual countries, which then decide whether or not to share their results • NGOs likely to play a larger supporting role in providing imagery expertise	• Norms based on a formal agreement among participants • Monitoring strategy depends on multinational coordination and the contributions of participating countries and NGOs • Imagery analysis provided by a small multinational cadre of imagery analysts and experts
Advantages	• Fewer obstacles to initiate and sustain • NGOs can play a more important role in initiating and supporting the transparency regime	• Greater opportunities for sharing monitoring costs and imagery analysis expertise • Potentially higher credibility for imagery analysis findings
Disadvantages	• More ambiguity regarding the regime's underlying norms • States will be unable to share the cost burdens of imagery collections and analysis • Less coherent monitoring strategy as states focus on particular areas of interest	• More difficult to negotiate and operate • Operations rely on receiving regular funding contributions from participating states •Coordination problems related to formulating an agreed-upon monitoring strategy

ticipants. The communication network (probably based on a combination of voice, fax, express mail, and e-mail capabilities) would link the appropriate points of contact in different governments, multinational organizations, and even NGOs, so that they can freely share monitoring information on the Spratly Islands, including commercial imagery scenes. Imagery data files might be shared through express mail deliveries or even an Internet distribution system so that participating governments and other entities could have their own experts review the data.

The specific nature of the monitoring strategy would depend on whether the transparency regime for the Spratly Islands is based on a formal agreement or is taking form through informal practices. In the first case, a higher degree of coordination and consensus among the governments participating in the transparency regime is required for efficiently collecting commercial satellite imagery to support the two main monitoring missions: 1) detecting whether any new locations for continuous human presence are under construction, and 2) whether the facilities at occupied islands and reefs are being substantially expanded or improved. In contrast, an informal approach is likely to offer only *ad hoc* coverage of key Spratly Islands locations based on the voluntary cooperation by certain governments or even NGOs with an ability to purchase and analyze satellite imagery.

Thus, the communications network provides a basis for coordinating the monitoring strategy in terms of collecting commercial satellite imagery of the Spratly Islands area. It would also permit a collective review of the analysis of new imagery data among the participants in the transparency regime to help develop some degree of consensus over what the imagery analysis reveals.

CONCLUSIONS

The new generation of commercial and civilian observation satellites offers a preferable means for monitoring the potentially escalatory developments of the competing claimants for the Spratlys. Unlike military reconnaissance overflights or nearby naval patrols, they can collect the necessary data on what is occurring without increasing the risk of provoking a forcible response, particularly during times of diplomatic tensions and military uncertainties.

However, the South China Sea littoral states not only have a common interest in avoiding armed conflict over the disputed Spratly Islands, but they also have shared interests in dealing with piracy, responding to natural disasters, undertaking environmental monitoring, and pursuing scientific research on ocean weather patterns. Multinational cooperation is essential for effectively using Earth observation systems to deal with these areas because a broad range of space-based sensors is relevant to each of these remote sensing missions. No single country possesses the full range of technical capabilities and expertise necessary for acquiring, processing, or analyzing the relevant imagery data from the diverse array of civilian and commercial observation satellites. Hence, a cooperative monitoring regime could greatly improve the chances that the countries bordering the South China Sea will obtain the information they need from imaging satellites when they most need the data. Encouraging cooperative monitoring in these areas of common interests might also serve to reduce some of the reluctance of various governments toward sharing satellite imagery data and other information necessary for effectively monitoring the disputed Spratly Islands.

NOTES

1. Antonia Handler Chayes and Abram Chayes, "Regime Architecture: Elements and Principles," in Janne E. Nolan, ed., *Global Engagement: Cooperation and Security in the 21st Century* (Washington, D.C.: The Brookings Institution, 1994), 65–68, 81–84.

2. Of course, an agreed interpretation of the norm could be developed that permits countries to undertake limited upgrades for safety and habitability at locations under their control among the Spratly Islands as long as prenotification is given. Any suspicions that improvements are aimed at making militarily significant upgrades could be addressed by allowing foreign observers to visit the island or reef in question for an on-site inspection once work on the planned change is underway or completed.

3. Mark J. Valencia, *China and the South China Sea Disputes*, Adelphi Papers (London, U.K.: International Institute for Strategic Studies, 1995), 8.

4. During the tensions over the Chinese expansion of the facilities at Mischief Reef in 1998, reconnaissance aircraft from the Philippines initially made close approaches to the disputed sites and later were order by Manila to remain at least 1,500 feet from the Chinese-occupied sites. Nonetheless, during the period of heightened tensions, observation aircraft reportedly approached the Mischief Reef facilities within a few hundred feet on several occasions to obtain close-up photographs of Chinese activities. In one instance, a U.S. congressmen was flown around Mischief Reef site in a Philippines military C-130 aircraft in mid-December for direct observation of the contested site. Bill Gertz, "China Makes Move on Spratlys," *Washington Times* (January 4, 1999), A1.

5. Li Hui, "Science Overrides Politics for East Asian Monsoon Study," *Science* (April 17, 1998), 373; and "International Team Links Monsoon to Sea Patterns," *Space News* (June 21, 1999), 20. Also see the Web site: http://climate.gsfc.nasa.gov/~kim/relacs/campaign/index.html/.

6. Ibid.

7. Alan Dupont, *The Environment and Security in Pacific Asia*, Adelphi Paper 319 (London, U.K.: International Institute for Strategic Studies, June 1998), 50–57.

8. "Red Tides Devastate Hong Kong fisheries," SeaWeb OceanUpdate (June 1998), Web site: http://www.seaweb.org/18update/hk.thml.

9. "U.S. Official Cites Severe Bleaching of Coral Reefs," report of Rafe Pomerance, deputy assistant Secretary of State for Environment and Development in report to the U.S. Coral Reef Task Force (March 10, 1999), Web site: www.usia.gov/curr.../washfile/newsitem.shtml; and T. J. Goreau and R. L Hayes, "Coral Bleaching and Ocean Hot Spots," *Ambio* 23 (1994), 176–180.

10. *Reefs at Risk: A Map-Based Indicator of Threats to the World's Coral Reefs* (Washington, D.C.: World Resources Institute, 1998); and Douglas H. Chadwick, "Coral in Peril," *National Geographic* 195 (January 1999), 34–37.

11. Ben Iannotta, "AVHRR Data May Help Predict Coral Reef Trouble," *Space News* (November 23–20, 1998), 18.

12. The SeaWiFS sensor uses six visible and two near infrared imaging channels to produce daily color images of the world's ocean that can indicate phytoplankton levels. It was built with funding from NASA and operates on the Oribital Sciences Corporation's Seatrack spacecraft. High-quality ocean color images provided by this sensor are being used for commercial purposes, such as "fishing" maps of phytoplankton concentrations for use by ocean fleets, as well as for scientific studies of environmental trends involving ocean temperatures, such as the *El Niño* effect. Presentation by Charles McClain, Manager, SeaWiFS Project Manager, "Sea WiFS Observations of Marine Ecosystems" at the "Action at the Interfaces of the Earth Sciences" public lecture series at the NASA Goddard Space Flight Center (June 8, 1999).

13. Ben Iannotta, "California Wants Radarsat's Eye on Shipping Lanes," *Space News* (November 9–15, 1998), 26.

14. A good survey of these imaging satellites with coastal resource applications can be found in National Ocean Service, *Remote Sensing for Coastal Resource Managers: An Overview* (Washington, D.C.: National Oceanic and Atmospheric Administration, U.S. Department of Commerce, February 1997), 29–77.

15. Peter J. Mumby, Peter S. Raines, David A. Gray, and Janet P. Gibson, "Geographic Information Systems: A Tool for Integrated Coastal Zone Management in Belize," *Coastal Management* 23 (1995), 111–121.

16. Mt. Pinatubo had more than a major local effect on Central Luzon by burying several low-lying villages and causing major population dislocations. It also had a significant diplomatic effect by compelling the United States to abandon its large air force presence at Clark Air

Force Base, and posed a major regional hazard to air traffic from its high-altitude ashfalls that could clog jet engines. The Philippines is monitoring another volcano, Mayon, which could produce a major eruption. Michael Sadim, "Perfect Cone, Active Dome," *The Manila Times* (July 5, 1999).

17. Disaster Information Task Force Report, *Harnessing Information and Technology for Disaster Management: The Global Disaster Information Network (GDIN),* (November 1997), 35–43.

18. Radarsat International (RSI) of Canada has helped Vietnam to develop an operational monitoring system for tracking floods and storms that makes use of RADARSAT imagery, flood data, and change detection software to support flood prediction, response, monitoring, and damage assessment. "Imagery Aids Vietnamese Disaster Management," *Imaging Notes* 15 (March/April 2000), 8.

19. Michael A. Taverna, "Europeans Poised to Begin Work on Global Monitoring Network," *Aviation Week and Space Technology* (October 23, 2000), 27–38.

20. William M. Carpenter and David G. Wiencek, "Maritime Piracy in Asia," in William M. Carpenter and David G. Wiencek, eds., *Asian Security Handbook 2000* (Armonk, NY: M.E. Sharpe, 2000), 88–98; and Kevin Sullivan and Mary Jordan, "High-Tech Pirates Ravage Asian Seas," *Washington Post* (July 5, 1999), A18.

21. Foreign Broadcast Information Service (FBIS), "Security Expert: PRC Military Commit Piracy in Spratlys," *Manila Business World* (May 13, 1999) (FBIS-EAS-1999-0513). In the mid-1990s, there was an upsurge of "quasi-military" incidents of piracy in the South China and East China Seas, which raised concern that such activities were condoned or supported by Chinese officials as an exercise of extra-territorial sovereignty. See Dr. Stanley B. Weeks, "Piracy in East Asia: Trends and Counter-Measures," paper presented at the 11th Asia-Pacific Roundtable (Kuala Lumpur, June 5–8, 1997).

22. Our estimate of the essential capabilities needed by the Imagery Analysis Staff would include two computing workstations (for example, UNIX or SPARC stations), image processing software, some equipment needed for imagery interpretation (for example, a scanner system, light table, and a stereoscope system), and some miscellaneous support equipment (for example, a color printer, a high capacity modem for sending and receiving large imagery data files, and a digital and hard copy storage system).

9

Environmental Remote Sensing of the South China Sea

Douglas O. Fuller

INTRODUCTION

Remote sensing of the seas provides one of the most efficient means to obtain timely information for monitoring and evaluation of marine resources. As a region characterized by geopolitical tensions, a history of piracy, heavy resource use, and population pressure, the South China Sea and environs represent an excellent case for space-based observation of environmental change. The pressures on the resource base of the South China Sea and surrounding coastal areas are myriad. So, too, is the number of satellite sensors and platforms that provide valuable information on the diminishing resources of the South China Sea itself.

Like most marine areas around the world, the South China Sea is experiencing rapid changes in habitat quality that are likely to damage fisheries and other aquatic life. Interactions between human activity on the land and sea are important but seldom studied. Human activity is becoming increasingly concentrated in the coastal areas and thus creating more environmental impacts that may be detected for many miles up- or downstream. The development of coastal aquaculture ponds in Peninsular Malaysia, for example, reduces the amount of mangrove vegetation, which serves as nurseries and breeding grounds for ocean fish. The mangroves also serve as valuable buffers against typhoons, which often strike the northern portion of the region with legendary ferocity during the southwest monsoon.[1] The mangrove forests, aquaculture ponds, typhoons, and coastal zone erosion are all visible from space, as are countless other features that make up the Sea's complex environment. While remote sensing of the South China Sea holds significant promise, the number of studies employing remote sensing there remains modest relative to Europe and North America, where expertise and data are widely available. The goal of this chapter is to review the principal environmental remote sensing studies that apply to the South China Sea and to discuss the use of past, present, and future satellite systems for monitoring the regional environmental resources. Before presenting relevant applications, it is worthwhile to review the physical and biological environments as well as some of the major environmental issues facing the South China Sea so that the potential and limitations of marine remote sensing can be fully appreciated.

THE PHYSICAL AND BIOLOGICAL SETTING

Covering an area of 3.4 million square kilometers, the South China Sea is the largest body of water in an archipelagic region situated between the Pacific and Indian Oceans. The depth of the South China Sea varies greatly from approximately 100 meters along the Sunda Shelf in the south to deep basins of around 4,000 meters in the north between Luzon (the Philippines) and China.[2] The Sea itself is the product of millions of years of tectonic activity, which is manifested today in the volcanoes that are found around portions of the area. According to B. Yao, multiple oceanic plate movements in the late Eocene through Early Oligocene resulted in the opening of the South China Sea basin from northwest to southeast. Like the Atlantic Basin, the sea floor of the South China Sea is still spreading.[3]

Ancient accumulations of organic sediments have produced high concentrations of hydrocarbon deposits in some areas. For example, it is believed that the Spratly Islands may contain up to 25 billion cubic meters of natural gas and 105 million barrels of crude oil.[4] Major gas finds have also been confirmed off the coast of Vietnam[5] and in the northeastern Beibu Gulf Basin of China[6] with corresponding increases in hydrocarbon exploration and ship traffic. Although the submarine gas and oil deposits are not readily detectable with satellite sensors, oil spills and ships (both wakes and superstructure) may be easily seen on radar and optical images.[7]

The geology of the Sea basin is also characterized by a high number of coral atolls, which are ring-shaped structures that are associated generally with eroded islands. Most of the coral reefs are based on the Tertiary strata, underlain by the micro-continental blocks, making them different from other oceanic atolls, which are mainly based on the volcanic seamounts. Guozhong recognizes three main types of atolls in the South China Sea—large atolls that consist of a multiple lagoon surrounded by separate table reefs or islets; medium atolls consisting of a ring-form single lagoon, surrounded by a continuous reef wall; and small atolls with a closed shallow water lagoon or isolated table reefs.[8] Although not as extensive and diverse as the coral reefs of the nearby Sulu and Sulawesi Seas, the South China Sea is the home of fairly rich coral-reef diversity, particularly along the coasts of Sabah, Malaysia, and Palawan.[9]

Limits to coral-reef diversity in the South China Sea may be explained by the sediment-laden rivers that flow into the Sea from mainland East Asia (the Pearl and Mekong, for example) and from the coast of Sarawak, Malaysia.[10] Sediment inputs from these rivers total approximately 7.2 billion tons annually, which increase turbidity such that the symbiotic photosynthetic algae (that is, zooxanthellae) of coral organisms cannot survive in many coastal areas. Increased runoff and sedimentation caused by development on surrounding land areas also pose a major threat to the living coral organisms of the region. For example, clear-cutting of tropical forest in Sabah, Malaysia, may produce soil erosion rates of up to 50,000 metric tons per square kilometer per year.[11] Much of this eroded soil makes its way downstream via rivers and streams where it is deposited along the coastline and eventually smothers the reefs.[12] A recent report by L. M. Chou mentions that coral cover in the Spratly Islands is relatively high due to the clear waters in the middle of the Sea but that recent development of a resort and airstrip on Swallow Reef has caused a massive loss of 75 percent coral cover.[13] Other examples of declines of coral reefs in the Indo-Pacific region from sediment and other forms of pollution are reviewed by C. R. Wilkinson[14] and D. Bryant *et al.*[15]

Biologically, the South China Sea belongs to the Pacific Coastal Biome and covers both the Sunda-Arafura Shelves Biogeographic Province and China Sea Coastal Biogeographic Province.[16] As A. Longhurst has argued, scientific knowledge of the distribution of marine organisms is poor relative to terrestrial ones, and therefore biological stratification of the seas is best accomplished using biophysical parameters such as primary production, turbidity or photic depth, salinity, upwelling zones, temperature, and seasonality.[17] Fortunately, many of these physical parameters may be obtained from satellite observations. Photic depth, or the depth to which light can penetrate open waters, provides a useful indicator of the potential productivity and biological diversity of the seas. Benthic communities within the euphotic zone, such as coral reefs and sea-grass beds, are among the most productive and diverse of all ecosystems.[18] Due to the relatively high sediment loads, the photic depth of the South China Sea is relatively shallow and generally ranges between 20–30 meters; whereas, in the coastal waters off eastern Australia and western South America, photic depth tends to vary between 40–50 meters.[19] Nevertheless, the biological productivity of the South China Sea is significant if one considers the number of commercial fishing grounds (63 compared to only 25 in the East China Sea) and the annual fish catch in China's seas. Figure 9.1 suggests that fishing pressure on the South China Sea has grown rapidly relative to the other coastal seas of China, which now ranks first among all countries in total fish catch.[20]

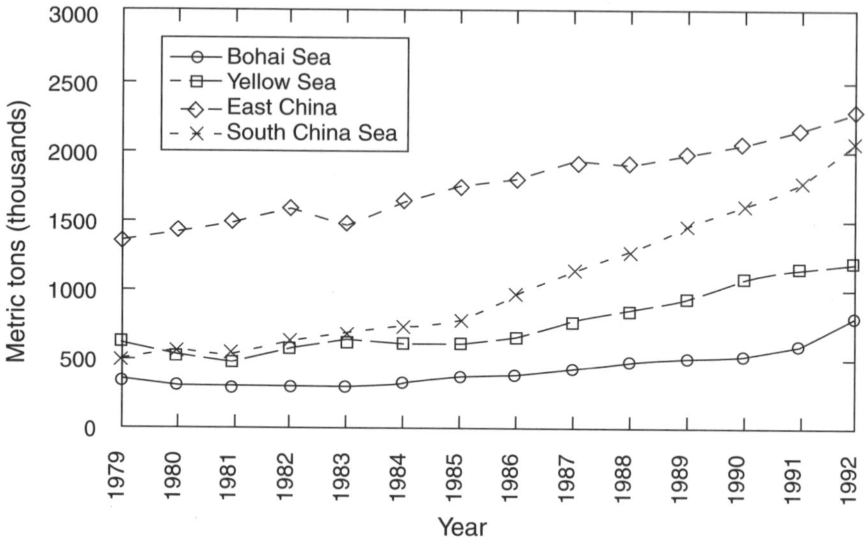

Figure 9.1

Size of the Annual Fish Catch Reported by China in Chinese Coastal Waters (measured in thousands of metric tons per year). Since 1984, the South China Sea has become an increasingly important source of China's annual fish catch and provides an indicator of the Sea's overall biological productivity.

Source: Web site: http://www.lib.noaa.gov/china/statistics.htm.

Although it is impossible to monitor the fish stocks directly with remote sensing, it is possible to monitor fishing fleets, water quality, and some benthic properties of fishing grounds in shallow areas with satellite sensors (discussed in later text). Particularly disturbing to the environmental community are destructive fishing practices in which cyanide is used around coral reefs to stun fish for the aquarium trade. Dynamite fishing and use of cyanide are prevalent in South China Sea, particularly in the Paracel Islands and around many parts of the Philippines, although steps are being taken to educate local fishermen about the dangers and to introduce more benign fishing methods such as fine-mesh nets.[21]

One of the most pressing environmental issues facing the South China Sea as well as other tropical seas is extensive coral bleaching in which reef-building coral organisms lose or expel their symbiotic zooxanthellae in massive bleaching events that occur over several months.[22] Elevated sea-surface temperatures (SST) relate closely to spatial patterns observed in bleaching events, and therefore these appear to be a major cause of bleaching, although the causal mechanisms are not yet fully understood.[23] The southeast Asian seas did not experience coral bleaching as extensively as the reefs found in the Indian Ocean during the last major El Niño event of 1997–1998, possibly because several major cool-water upwelling zones exist in that region. A combination of anecdotal and scientific evidence suggests that bleaching is reversible in some coral species and that recovery may be underway in parts of the affected areas. However, experts generally agree that bleaching and mortality will continue to increase worldwide as sea-surface temperatures and sea levels continue to rise.[24] Because living corals are found in the euphotic zone—where sufficient light exists for photosynthesis and bleaching produces a change in the spectral reflectance of reefs—remote sensing has an important potential role to play in mapping and monitoring this serious environmental problem.[25]

SATELLITE REMOTE SENSING OF SURFACE WATERS

One of the greatest challenges in environmental remote sensing is to obtain information on water quality, column constituents, and bottom properties using space-borne sensors. Radiance that reaches a sensor over open water may originate from many sources, including atmospheric path radiance, surface scattering, volume scattering from the water column, upwelling radiance from suspended matter such as phytoplankton and sediment (that is, the so-called "yellow" substances), and the radiance reflected from the bottom substrate. The signal over water bodies is generally quite weak in comparison to that over land and the satellite systems designed for marine applications generally require enhanced radiometric sensitivity. In a recent review of the remote sensing of coral reefs, Holden and LeDrew identified atmospheric path radiance, specular reflectance (sunglint) off the water surface, and separation of water column signals from the bottom (substrate signal) as some of the major challenges that remain in marine remote sensing.[26] In addition, the geo-location of small oceanic features such as coral reefs is complicated by a lack of reliable navigational charts or other maps produced at suitable scales. Lack of reliable data on sea-floor depth (bathymetry) also complicates assessment of the origins of the satellite signal over marine environments of variable depth.[27]

Despite these technical challenges, scientists have made considerable advances in marine observation using optical and thermal bands on coarse-resolution systems such as the Coastal Zone Color Scanner (CZCS) and the

Advanced Very High Resolution Radiometer (AVHRR). CZCS imagery has made a particularly valuable contribution to our understanding of ocean productivity (see Table 9.1). This sensor operated continuously from 1978–1986 and produced a rich archive of imagery on ocean color and productivity that has been the subject of many studies on global carbon balance and phytoplankton production.

Sea-surface temperature (SST) is one of the most important factors controlling storm development and intensity, and thermal infrared imagery from satellites has provided many insights on how SST varies in the South China Sea. Thermal imagery provided by geostationary systems such as the Geostationary Operational Environmental Satellite (GOES) and the

Table 9.1

Major Satellite Sensors for Monitoring the Marine Environment

Sensor	*Band*	*Band center (nm)*	*Bandwidth (nm)*	*Major utility*
CZCS	1	443	423–463	chlorophyll a
	2	520	500–520	chlorophyll a
	3	550	520–570	yellow substances
	4	670	650–690	chlorophyll a
	5	750	650–850	chlorophyll a suspended sediment
	6	11,500	9,500–13,500	sea-surface temperature
SeaWiFS	1	412	402–422	yellow substances
	2	443	433–453	chlorophyll a
	3	490	480–500	chlorophyll a
	4	510	500–520	chlorophyll a
	5	555	545–565	yellow substances
	6	670	660–680	chlorophyll a
	7	765	745–785	atmospheric correction
	8	865	845–885	atmospheric correction

Source: Adapted from J. R. Jensen, *Remote Sensing of the Environment: An Earth Resources Perspective* (Upper Saddle River, NJ: Prentice Hall, 2000).

AVHRR also has provided important data on the location and temporal dynamics of upwelling zones and changes in sea-surface temperature, including recent warming events associated with El Niño. In addition, synthetic aperture radar (SAR) systems like the European Research Satellites (ERS) 1 and 2, the Japan Environmental Research Satellite (JERS), and Radarsat have produced valuable insights on wave heights, wind speeds, ship wakes, and other sea-surface manifestations. In August 1997, the National Aeronautics and Space Administration (NASA) and OrbImage, Inc. launched the Sea-viewing Wide Field of View Sensor (SeaWiFS), which was designed specifically for monitoring ocean color and productivity. Like the CZCS, SeaWiFS consists of a small number of narrow, discrete bands covering the visible to near-infrared (NIR) wavelength range at approximately 1-km spatial resolution (refer to Table 9.1). As J. R. Jensen points out, although SeaWiFS provides a valuable new source of data on ocean color, it lacks a band around 700 nm, which would provide data on the type and concentration of inorganic suspended matter in surface waters.[28]

Marine biologists are accustomed to working at relatively fine spatial scales and therefore are particularly interested in new satellite systems that can provide spatial and spectral resolution much higher than the older systems such as CZCS and the AVHRR.[29] For this reason, many researchers have used moderate spatial resolution systems such as Landsat Thematic Mapper (TM) and SPOT for a variety of site-specific mapping applications. TM imagery is often preferred over SPOT because the former possesses a blue band that can measure water-leaving radiance from the deepest portions of the euphotic zone. To map tropical coastal environments greater than 60 km in any direction with coarse detail, Mumby *et al.* found that Landsat TM was the most accurate and cost-effective satellite sensor (or SPOT XS when <60 km).[30] The authors found that Landsat Multi Spectral Scanner (MSS) was consistently the least accurate sensor among moderate resolution systems. However, maps from Compact Airborne Spectrographic Imager (CASI) were significantly more accurate than the satellite sensors and aerial photographs. Overall, the authors extracted detailed habitat information (< nine reef classes) with a maximum accuracy of 37 percent when based on satellite imagery, although aerial photography and CASI achieved accuracies of 67 and 81 percent, respectively.[31]

After reviewing the use of moderate resolution sensors for mapping benthic communities, Holden and LeDrew concluded that consistent confusion among certain classes confounds accurate mapping of shallow areas (<30 meters) with moderate-resolution multispectral systems.[32] For example, submerged aquatic vegetation and coral reefs are often confused in satellite-based classifications as are substrate and suspended matter. Moreover, studies that aim to map benthic environments usually assume homogeneous, clear waters throughout the image,[33] which may be unrealistic in some areas. A large portion of the error in mapping benthic areas may be caused as well by the patchiness of marine habitats such as sea-grasses and corals, which may be juxtaposed within a single 30-meter pixel of a TM image.[34] Although intended primarily for mapping urban areas and infrastructure, very high-resolution (for example, four meters) commercial systems such as IKONOS and OrbView-3, which possess visible/NIR bands similar to TM[35] may provide marine biologists with a new data set for mapping important coastal and benthic communities.

Studies by Mumby *et al.*[36] and Clark *et al.*[37] suggest that use of imaging spectrometers may help overcome some of the difficulties of mapping marine environments posed by moderate-resolution multispectral systems. The Medium Resolution Imaging Spectrometer Instrument (MERIS) is one such system that is due to be launched in July 2001 on the ENVISAT-1 satellite of the European Space Agency[38] and will allow global coverage of the Earth in three days. MERIS is a 68.5° field-of-view, pushbroom imaging spectrometer that will provide imagery of the marine environment at a ground spatial resolution of 300 meters in 15 spectral bands. The primary mission of MERIS is the measurement of color in the oceans and in coastal areas.[39] As with CZCS and SeaWiFS, ocean color measurements can be converted into chlorophyll pigment and suspended sediment concentration over the marine domain.[40] High spectral resolution (that is, narrow, contiguous bands) is also achieved by the Moderate Resolution Imaging Spectrometer (MODIS), which was launched by NASA in December 1999 as one of the flagship instruments on the Terra satellite platform. MODIS consists of 36 spectral bands from 620 nm to 14,385 nm, and spatial resolution ranges from 250 meters for Bands 1 and 2 (620–670 nm and 841–876 nm, respectively) to 1 km for most other visible and infrared bands. Although more sensitive to sun-glint because of its wide scan angle range (±55°), MODIS has already shown promise in ocean color mapping as evidenced by NASA's early MODIS ocean-color products.

REMOTE SENSING STUDIES OF THE SOUTH CHINA SEA ENVIRONMENT

In workshop proceedings on the South China Sea environment published by the Asian Development Bank in 1995, one participant commented that "To date, no large-scale spill has been reported in the South China Sea."[41] Thanks to recent SAR-based studies conducted with ERS and Radarsat data, however, we now know that oil pollution from shipping operations in and around the South China Sea constitutes one of the major threats to the environment in the region.[42] This example illustrates how satellite remote sensing can rapidly change our perception of the environment as well as the importance of remote sensing facilities and personnel for data capture and analysis. At least eleven remote sensing centers in the region possess the ability to gather or process satellite data covering the South China Sea and environs, several of which are cited in Table 9.2. In addition, the severe forest fires in Indonesia over the past several years have prompted the establishment of a number of new remote sensing projects, many of which established AVHRR receiving stations to map fires in real time.[43] Although dealing explicitly with a land-based issue, these Indonesian fire-monitoring projects have produced a great deal of imagery that covers the southern portion of the Sea and have revealed the extent of smoke and haze blanketing the region. A haze map produced by Hardiono[44] based on the Total Ozone Mapping Spectrometer (TOMS) aerosol index[45] revealed moderate to heavy aerosols from the fires as far north as the Gulf of Thailand. A rough estimate produced in a recent report by Barber and Schweithelm suggests that the choking blanket of smoke and haze affected some 20 million people over a period of several months during the 1997–1998 El Niño.[46]

Of the remote sensing centers currently operating in the region, the Centre for Remote Imaging, Sensing and Processing (CRISP) at the National University of Singapore has carried out some of the most intensive monitoring of the marine environment of Southeast Asia. For example, CRISP researchers have collected and analyzed 5,029 ERS SAR scenes covering southeast Asian waters to map oil slick occurrence and intensity from 1995–1998. Lu reports that CRISP detected a total of 10,988 oil slicks in 2,292 scenes and found that 80 percent of the total study area (extending from the eastern Indian Ocean to the Sulawesi Sea) had been polluted at least once by an oil slick. Of the areas examined with ERS SAR imagery, the South China Sea, Gulf of Thailand, and Straits of Malacca showed the greatest intensity of oil slicks with greater than 25 per 10,000 square km. Similarly, CRISP researchers have used SPOT imagery in conjunction with ERS and Radarsat to track ships around Singapore that are responsible for oil spills.

Monitoring of individual ships has allowed CRISP researchers to produce evidence to Singapore legal authorities which led to the conviction of the officers from an offending vessel with fines totaling $1.25 million Singapore and a jail term for the ship's captain.[47]

Radar imagery has also been used to study the origin and occurrence of waves in the South China Sea. For example, Liu *et al.* used ERS-1 data to study the characteristics of internal waves northeast and south of Taiwan in the East China Sea, and east of Hainan Island in the South China Sea. The authors reported sets of stable, isolated waves propagating shoreward from the edge of the continental shelf that were observed in the SAR images. On the basis of the assumption of a semidiurnal tidal origin, the wave speed was estimated to be consistent with internal wave theory. By using the SAR images and hydrographic data, internal waves were identified in shallow water by a thicker mixed layer as compared with the bottom layer on the continental shelf.[48]

Table 9.2

Major Remote Sensing Centers in the South China Sea Region

Center	*Principal Satellite Data*	*Internet URL*	*Major Applications*
Australian Center for Remote Sensing	Landsat, SPOT, ERS, JERS, NOAA	http://www.auslig.gov.au/acres/	data capture/processing
Centre for Remote Imaging, Sensing and Processing (CRISP), Singapore	SPOT, ERS, Radarsat, NOAA	http://www.crisp.nus.edu.sg/	data capture/processing marine environment, land cover, forest fires
Center for Space and Remote Sensing Research, Taiwan	NOAA, SPOT	http://www.csrsr.ncu.edu.tw/english.ver	weather, fires, vegetation conditions, land cover, sea surface temperature
Indonesian Remote Sensing Agency—LAPAN	NOAA, Landsat	http://www.lapanrs.com/	data capture/processing, land cover, fires
Institute of Remote Sensing Applications (IRSA) of the National Remote Sensing Centre, China	NOAA, SAR	http://www.friends-partners.org/~china/ins/IRSA/irsa.html	geologic mapping, mineral exploration, land resources, crop monitoring
Malaysian Centre for Remote Sensing	NOAA-AVHRR, SPOT, Landsat	http://www.macres.gov.my/	land cover mapping, environmental monitoring
Space Technology and Applications Research Programme, Asian Institute of Technology, Thailand	TM, SPOT, AVHRR, JERS, ERS	http://www.star.ait.ac.th/	coastal zone management land cover mapping GIS
Asian Centre for Research on Remote Sensing, AIT, Thailand	ERS, JERS, NOAA-AVHRR	http://www.acrors.ait.ac.th/	data capture/processing SST, wind and wave heights, coastal zone management

Sources: Compiled from various Internet sites. See, for example, http://www.cz3.nus.edu.sg.

Radar altimetry has also been used to study wave heights and wind speeds in order to understand the development and movement of storms in the South China Sea. Based on the altimetric observations over the South China Sea from the Geosat satellite during the period between November 1986 to March 1989, Qi *et al.* determined the statistical characteristics of the sea surface wind speeds and the seasonal features of the wind field. The results show that the distributions of the seasonal averaged wind speeds manifest the signatures of the major weather processes, such as monsoons, typhoons, and subtropical highs. Data from satellite altimetry also showed that the wind fields exhibited a southward-increasing trend in the summer and a southward-decreasing trend in the other three seasons. The analysis of Geosat altimeter data also revealed a zone of strong winds situated at 10 degree N, 110 degree E throughout the year, with a slight seasonal variation in the center position and range.[49]

Chu *et al.* examined sea surface elevation in the South China Sea using the Topex/Poseidon altimeter data from 1992 to 1995. The data revealed significant annual variations in the sea level. In winter, sea levels tended to be low over the entire deep basin with two local lows centered off Luzon and the Sunda Shelf. In summer, sea level was high off Luzon and the Sunda Shelf, with a low off Vietnam separating the two highs. The boundary between the Vietnam low and Sunda high coincides with the location of a jet stream leaving the coast of Vietnam.[50] Such variations in sea surface height are important not only because they affect erosion in the coastal areas but also because reefs and other submerged ecosystems may be damaged by extreme high and low water levels.

Nearly 70 percent of the population of Southeast Asia lives within 70 km of the coastline,[51] therefore predicting storm intensity and its precursors is of utmost importance. Using National Oceanic and Atmospheric Administration (NOAA)-AVHRR and Landsat data, D. Ke *et al.* studied the frequencies of cyclones in the Yellow Sea, the East China and South China Sea, and their relationship to surface temperatures of the Tropical Pacific. The authors found that the number of cyclones during El Niño decreased and that strong thermoclines and haloclines appeared in the South China Sea coastal area.[52] This early satellite-based study helped to confirm the influence of El Niño on the climate and biogeochemistry in the Southeast Asian seas.

The magnitude of El Niño in 1997–1998 was one of the strongest of the century. This event increased SST throughout the Indo-Pacific and was associated with widespread impacts on the world's tropical seas.[53] One notable impact of the SST increase in the coastal zones of the South China Sea is an increase in the number of harmful phytoplankton blooms, also known as "red tides." Yin *et al.* used NOAA-AVHRR data to study a series of red tides that occurred in Hong Kong territorial waters between mid-March and mid-April 1998 in which the major species was *Gyrodinium aureolum*. They used a time series of satellite images to track the development of the harmful algal bloom and relate its movement to physical oceanographic conditions. Prior to this red tide event, other harmful algal blooms had been reported (November 1997–February 1998) along the south China coast to the east of Hong Kong, although different species were involved. AVHRR images revealed a progression of harmful algal blooms from northeast to southwest that coincided with the southwesterly movement of the south China coastal current. The authors also employed images of chlorophyll derived from SeaWiFS sensor data to confirm high near-shore concentrations of this pigment in the same region. The SST images showed a warm tongue pointing north to the south China coast, which suggested that the warm water from the South China Sea might have been piling up toward the coastline. As a result, the warm coastal waters of the south China coast may have become trapped in the coastal zone. The authors speculate that the combination of warm trapped water and coastal-zone eutrophication from land sources (for example, fertilizers, sewage, and sediments) may produce a set of unique conditions favoring such harmful algal blooms during El Niño events.[54]

Global monitoring of sea conditions by satellites is also improving our understanding of the long-term thermal dynamics of the South China Sea. Worldwide satellite and buoy data show long-term warming of the world's tropical seas,[55] which suggests an increasing probability of future, positive SST anomalies. NOAA currently produces a twice weekly global SST chart depicting anomalies (or departures) from mean conditions. The product, which is known as the Bleaching Hotspot Chart, is based on long-term NOAA-AVHRR observations and is intended to alert authorities to potential warm areas in the tropical seas at risk of coral bleaching.[56] NOAA currently maintains an image archive depicting anomalies, although the data are resampled to a very coarse 36-km spatial resolution. This approach relies on a maximum monthly mean SST climatology as the input threshold for the Coral Bleaching Hotspot charts on which coral bleaching events have been noted in areas where the SST exceeds the threshold by 1°C or more. Although spatially coarse, monthly image data from 1984 to the present reveal that the South China Sea's SST varies as a function of El Niño events. The 1997–1998 El Niño produced the sixth major worldwide bleaching event since 1979[57] and NOAA imagery shows that January anomalies during the El Niño years of 1994–1995 and 1997–1998 were strongly positive in the South China Sea. Although few reports are available on the conditions of the reefs after the last event, the magnitude of the warming in the Sea was similar to that observed in the Indian Ocean, where massive bleaching was reported.[58] Relative to Micronesia, Polynesia, the Indian Ocean, and the Caribbean, few field reports appear to be available on the status of the reefs to confirm whether elevated temperatures in the South China Sea indeed produced extensive bleaching there.

CONCLUSIONS

The world's coastal and marine resources are clearly in a state of decline from overfishing, the destruction of coastal ecosystems, pollution, and climate change. The remote sensing and other studies previously cited suggest that the South China Sea is no exception in this regard. However, relative to other major seas of the world, the South China Sea has not been studied intensively with satellite remote sensing. Nevertheless, because the Sea is the subject of major territorial disputes[59] and access is restricted in many areas due to poor infrastructure, satellite observations provide the most comprehensive, consistent, and long-term way to monitor its environment.

Although remote sensing of the living resources such as coral reefs, seagrass beds, and other benthic communities remains a major technical challenge, high-spatial resolution and fine-spectral resolution sensors promise to fill some of the information gaps that remain in respect to the Sea's living resource base. SeaWiFS and NOAA-AVHRR already provide excellent sources of information on phytoplankton concentration and SST, respectively. More systematic use of these tools to study the interactions of long-term warming, El Niño events, harmful algal blooms, and coral bleaching is called for. Use of such data along with limited in-situ measurements could lead to the development of early warning systems, which can alert authorities to potentially harmful episodes of pollution from sediments, fertilizers, oil, and so forth. Several Coral Reef Early Warning Systems (CREWS) have been proposed for different parts of the world (see, for example, http://coral.aoml.noaa.gov/corvil/index.html), however, it remains unclear whether support will emerge from donor governments to establish such systems. Moreover, given the current emphasis on resolving geopolitical problems, it is uncertain that ASEAN or other regional bodies will take significant action based on the data they might someday produce. Meanwhile, decentralized remote sensing efforts based in the region will continue to provide timely information on how the South China Sea's environment is faring relative to the world's other great seas.

NOTES

1. J. Howes, "Mangroves" in S. Sani, ed., *The Encyclopedia of Malaysia: Volume 1: The Environment* (Singapore: Archipelago Press, 1998), 56–57.
2. C. L. Ming and P. M. Alino, *An Underwater Guide to the South China Sea* (Singapore: Times Editions, 1992), 144.
3. B. Yao, "Tectonic Evolution of the South China Sea in Cenozoic," *Marine Geology and Quaternary Geology/Haiyang Dizhi Yu Disiji Dizhi, Quingdao*, 16 (1996), 1–13.

4. W. Scholes and B. S. Yurchak, "Spratly Islands Stand-off in The South China Sea: A Radar Study of the Eye of Tropical Cyclone Irving During its Passage Over Tonkin Gulf," *Petroleum Review* 51 (1997), 38.

5. T. Knott and H. Olson, "Oil May Lie on Troubled Waters: Marine Traffic in The South China Sea," *Petroleum Review* 50 (1996), 236–237.

6. T. Xie and T. Zhang, "The Petroleum System and Exploration Targets in the Northeastern Part of Beibu Gulf Basin China Offshore Oil and Gas," *Geology/Zhongguo Haishang Youqi. Dizhi [China Offshore Oil Gas [GEOL.]/Zhongguo Haishang Youqi [Dizhi]]* 10 (1996), 297–303.

7. J. Lu, "Marine Oil Spill Surveillance and Mapping Using Remote Sensing in Singapore," *Integrated Coastal Zone Management* 1 (2000), 217–222.

8. W. Guozhong, "The Coral Atolls in the South China Sea and the Comparison with the World Atolls," *Abstracts of Papers Presented at the 30th International Geological Congress* (1996), 204.

9. L. M. Chou, "Status of Southeast Asian Coral Reefs," in C. Wilkinson, ed., *The Status of Coral Reefs of the World: 1998* (Cape Fergunson, Queensland: Australian Institute of Marine Science, 1998), 79–87.

10. Ibid.

11. S. M. S. Abdullah, "Soil Erosion," in S. Sani, ed., *The Encyclopedia of Malaysia: Volume 1: The Environment* (Singapore: Archipelago Press, 1998), 26–27.

12. D. Bryant, L. Burke, J. McManus, and M. Spalding, *Reefs at Risk: A Map-based Indicator of Threats to the World's Coral Reefs* (Washington, D.C.: World Resources Institute, 1998), p. 56.

13. L. M. Chou, "Status of Southeast Asian Coral Reefs," in C. Wilkinson, ed., *The Status of Coral Reefs of the World: 1998* (Cape Fergunson, Queensland: Australian Institute of Marine Science, 1998), 83.

14. C. R. Wilkinson, "Global and Local Threats to Coral Reef Functioning and Existence: Review and Predictions," *Marine and Freshwater Research* 50 (1999), 867–78.

15. D. Bryant, L. Burke, J. McManus, and M. Spalding, *Reefs at Risk: A Map-based Indicator of Threats to the World's Coral Reefs* (Washington, D.C.: World Resources Institute, 1998), 56.

16. A. Longhurst, *Ecological Geography of the Sea* (San Diego, CA: Academic Press, 1998).

17. Ibid.

18. J. T. O. Kirk, *Light and Photosynthesis in Aquatic Ecosystems,* 2nd ed. (Cambridge, U.K. : Cambridge University Press, 1994), 428–429; and O. Hoegh-Guldberg, "Climate Change, Coral Bleaching and the Future of the World's Coral Reefs," *Marine and Freshwater Research* 50 (1999), 839–866.

19. A. Longhurst, *Ecological Geography of the Sea* (San Diego, CA: Academic Press, 1998).

20. Japan Environmental Council, *The State of the Environment in Asia 1999/2000* (Tokyo: Springer-Verlag, 2000), 197.

21. C. V. Barber and V. R. Pratt, *Sullied Seas: Strategies for Combating Cyanide Fishing in Southeast Asia and Beyond* (Washington, D.C.: World Resources Institute and International Marine Alliance, 1997); and D. Bryant, L. Burke, J. McManus, and M. Spalding, *Reefs at Risk: A Map-based Indicator of Threats to the World's Coral Reefs* (Washington, D.C.: World Resources Institute, 1998).

22. O. Hoegh-Guldberg, "Climate Change, Coral Bleaching and the Future of the World's Coral Reefs." *Marine and Freshwater Research,* 50 (1999), 839–866.

23. Ibid.

24. O. Hoegh-Guldberg, "Climate Change, Coral Bleaching and the Future of the World's Coral Reefs," *Marine and Freshwater Research,* 50 (1999), 839–866; L. Stone, A. Huppert, B. Rajagopalan, H. Bhasin, and Y. Loya, "Mass Coral Reef Bleaching: A Recent Outcome of Increased El Niño Activity?" *Ecology Letters* 2 (1999), 325–330; and C. R. Wilkinson, "Global and Local Threats to Coral Reef Functioning and Existence: Review and Predictions," *Marine and Freshwater Research* 50 (1999), 873.

25. H. Holden and E. LeDrew, "Spectral Discrimination of Healthy and Non-healthy Corals Based on Cluster Analysis, Principal Components and Derivative Spectroscopy," *Remote sensing of Environment* 65 (1998), 217–224; and H. Holden and E. LeDrew, "The Scientific Issues Surrounding Remote Detection of Submerged Coral Ecosystems," *Progress in Physical Geography* 22 (1998), 190–221.

26. H. Holden and E. LeDrew, "The Scientific Issues Surrounding Remote Detection of Submerged Coral Ecosystems," *Progress in Physical Geography* 22 (1998), 190–221.

27. P. Bierworth, T. Lee, and R. Burne, "Shallow Sea Floor Reflectance and Water Depth Derived by Unmixing Multispectral Imagery," *Photogrammetric Engineering and Remote Sensing* 59 (1993), 331–338.

28. J. R. Jensen, *Remote Sensing of the Environment* (Upper Saddle River, NJ: Prentice Hall, 1999), 391.

29. USDOC/NOAA/NESDIS, NOAA/CSC, and ICLARM, *International Workshop On The Use Of Remote Sensing Tools For Mapping And Monitoring Coral Reefs* (Honolulu, Hawaii: East-West Center, June 7–10, 1999), proceedings available on Web site: http://coral.aoml.noaa.gov/corvil/index.html.

30. For maps with intermediate marine habitat detail, aerial photography exhibited similar accuracy to Landsat TM, SPOT XS, SPOT Pan, and merged Landsat TM/SPOT Pan.

31. P. J. Mumby, E. P. Green, A. J. Edwards, and C. D. Clark, "Coral Reef Habitat-Mapping: How Much Detail Can Remote Sensing Provide?" *Marine Biology* 130 (1997), 193–202.

32. H. Holden and E. LeDrew, "The Scientific Issues Surrounding Remote Detection of Submerged Coral Ecosystems," *Progress in Physical Geography* 22 (1998), 205.

33. P. Bierwoth, T. Lee, and R. Burne, "Shallow Sea Floor Reflectance and Water Depth Derived by Unmixing Multispectral Imagery," *Photogrammetric Engineering and Remote Sensing* 59 (1993), 333.

34. J. J. Luczkovich, T. W. Wagner, J. L. Michalek, and R. W. Stoffle, "Discrimination of Coral Reefs, Seagrass Meadows, and Sand Bottom Types from Space: a Dominican Republic Case Study," *Photogrammetric Engineering and Remote Sensing* 59 (1993), 385–389.

35. J. R. Jensen, *Remote Sensing of the Environment* (Upper Saddle River, NJ: Prentice Hall, 1999), 226.

36. P. J. Mumby, E. P. Green, A. J. Edwards, and C. D. Clark, "Coral Reef Habitat-Mapping: How Much Detail Can Remote Sensing Provide?" *Marine Biology* 130 (1997), 193.

37. C. Clark, H. Ripley, E. Green, A. Edwards, and P. J. Mumby, "Mapping and Measurement of Tropical Coastal Environments with Hyperspectral and High Spatial Resolution Data," *International Journal of Remote Sensing* 18 (1997), 237–242.

38. European Space Agency, ENVISAT 1, see http://envisat.estec.esa.nl/.

39. A. Bricaud, A. Morel, and V. Barale, "MERIS Potential for Ocean Colour Studies in the Open Ocean," *International Journal of Remote Sensing* 20 (1999), 1757–1769.

40. See http://envisat.estec.esa.nl/.

41. H. G. Tucker, "Marine Oil Pollution Control in the South China Sea Region," *Coastal and Marine Environmental Management*, Proceedings (Bangkok, Thailand: Asian Development Bank, Manila, 1995), 236–253.
42. J. Lu, "Marine Oil Spill Surveillance and Mapping Using Remote Sensing in Singapore," *Integrated Coastal Zone Management* 1 (2000), 217–222.
43. D. O. Fuller and M. Fulk, "Comparison of NOAA-AVHRR and DMSP-OLS for Operational Fire Monitoring in Kalimantan, Indonesia," *International Journal of Remote Sensing* 21 (2000), 181–187; D. O. Fuller, *Mapping the 1997 Fires in Indonesia: Data, Methods, and Future Perspectives* (Jakarta: World Wide Fund for Nature Indonesia, 2000); and R. Denis, *A Review of Fire Projects in Indonesia (1982–1998)* (Bogor, Indonesia: Centre for International Forestry Research, 1999), 105.
44. C. V. Barber and J. Schweithelm, *Trial by Fire: Forest Fires and Forestry Policy in Indonesia's Era of Crisis and Reform* (Washington, D.C.: World Resources Institute, 2000), 76.
45. T. Nakajima, A. Higurashi, N. Takeushi, and J. R. Herman, "Satellite and Ground-based Study of Optical Properties of 1997 Indonesian Forest Fire Aerosols," *Geophysical Research Letters* 26 (1999), 2421–2424.
46. J. Lash, "Foreword" in C. V. Barber and J. Schweithelm, *Trial by Fire: Forest Fires and Forestry Policy in Indonesia's Era of Crisis and Reform* (Washington, D.C.: World Resources Institute, 2000), vi.
47. J. Lu, "Marine Oil Spill Surveillance and Mapping Using Remote Sensing in Singapore," *Integrated Coastal Zone Management* 1 (2000), 217–222.
48. A. K. Liu, Y. S. Chang, Ming-K. Hsu, Nai K. Liang. "Evolution of Nonlinear Internal Waves in the East and South China Seas," *Journal of Geophysical Research* 103 (1998), 7995–8008.
49. Y. Qi, P. Shi, and Q. Mao, "Seasonal characteristics of the sea surface wind speeds in the South China Sea by remote sensing," *Tropic oceanology/Redai Haiyang Guangzhou* 15 (1996), 68–73.
50. P. C. Chu, Y. Chen, S. Lu, P.-T. Shaw, S.-Y. Chao, and L.-L. Fu, "Wind-driven South China Sea Deep Basin Warm-core/Cool-core Eddies: Sea Surface Height Variations in The South China Sea From Satellite Altimetry," *Journal of Oceanography* 54, 347–360.
51. W. Huang and M. Fang, "Monthly Composites of Sea Surface Temperature of the South China Sea," *International Journal of Remote Sensing* 19 (1998), 2435–2438.

52. D. Ke, M. I. Seeni Mohd, S. Ahmad, A. Abdullah, H. Kawamura, L. G. Cordner, and J. C. L. Chan, "The Influence of ENSO on the Marine Environment of Our Country: Satellite Remote Sensing for Marine Applications: Case Studies on Sea Surface Temperature and Sea Bottom Features Mapping, The Spratly Islands Dispute and the Law of the Sea, Prediction of the Interannual Variations of Tropical Cyclone Movement Over Regions of the Western North Pacific." *Marine forecasts/Haiyang Yubao* 11 (1994), 32–37.

53. C. Wilkinson, *Status of Coral Reefs of the World: 1998* (Cape Fergunson, Queensland: Australian Institute of Marine Science, 1998), 11.

54. K. Yin, P. J. Harrison, J. Chen, and W. Huang, "Red Tides During Spring 1998 in Hong Kong: Is El Niño Responsible?" *Marine Ecology Progress Series* 187 (1999), 289–294.

55. M. A. Cane, A. C. Clement, A. Kaplan, Y. Kushnir, D. Pozdnyakov, R. Seager, S. E. Zebiak, and R. Murtugudde, "Twentieth-century Sea Surface Temperature Trends," *Science* 275 (1997), 957–960.

56. T. Goreau, T. McClanahan, R. Hayes, and A. Strong, "Conservation of Coral Reefs After the 1998 Global Bleaching Event," *Conservation Biology* 14 (2000), 5–15.

57. O. Hoegh-Guldberg, "Climate change, coral bleaching and the future of the world's coral reefs," *Marine and Freshwater Research,* 50 (1999), 844.

58. C. Wilkinson, "The 1997–1998 Mass Bleaching Event Around the World," (Cape Fergunson, Queensland: Australian Institute of Marine Science, 1998), 15–38.

59. C. C. Joyner, "The Spratly Islands Dispute in The South China Sea," *Integrated Coastal Zone Management* 1 (2000), 85–89.

10

Contemporary Maritime Piracy in Southeast Asia

Peter Chalk

INTRODUCTION

Maritime piracy has been endemic to the Southeast Asian region for hundreds of years.[1] Indeed, the narratives of *Shih Fa-Hsien*, a Buddhist monk from Ceylon, record cases of raising, robbing, and other instance of marauding in the waters of the South China Sea as early as AD 414.[2] In his study on piracy, L. Wright shows how Dutch attempts to monopolize the spice trade through the Malacca Strait from the 1670s onward served to dramatically increase the instances of piracy in Southeast Asia by disrupting and distorting local trading patterns.[3] By the nineteenth century, marauding had become endemic to much of the Malayan world, largely because it was integral to the political dynamics of intertribal warfare and archipelagic empire building in the region.[4] Although superior technology, gunboats, and new, large steam-driven vessels did result in a decline in piracy by the latter part of the nineteenth century, it was never completely eradicated as a problem during the colonial period.[5]

Echoes of privateering continued well into the twentieth century. Sea raiding, thought to be condoned by Indonesia, was a factor that the Commonwealth naval forces had to take into account during the Malayan Emergency (1958–1960) and Sukarno's subsequent policy of Confrontation (*Konfrontasi*) against the newly created Malaysian Federation between 1963 and 1966.[6] Moreover, during the 1970s, numerous attacks against Vietnamese boat people fleeing the Communist advance in Indochina were documented, with many of the assaults apparently carried out by raiders based in Malaysia and Thailand.[7] Despite these instances, for much of the Twentieth century the issue of piracy was sidelined in importance to the wider East-West ideological struggle that was taking place between the superpowers and their respective blocs. However, with the end of that conflict and the somewhat artificial constraints that it placed on regional security, piracy is once again emerging as a prominent security problem in Southeast Asia. In 1999, for instance, the Malaysian-based Regional Piracy Center (RPC) documented a total of 160 attacks in the region, over half the global tally registered for the year and the highest figure currently on record.[8]

Concern over the growing seriousness posed by maritime violence and crime in Southeast Asia has prompted the international shipping industry to look at innovative ways of improving the security of vessels while traveling in piracy-prone regions. One potentially useful area that is now being emphasized is satellite and information technology. This chapter will examine the evolving nature of piracy in Southeast Asia and discuss some of the more important aerial/electronic imaging, tracking, and reporting systems that have been developed to help mitigate this particular threat.

WHAT IS PIRACY?

For the purposes of this chapter, piracy is defined as "any act of boarding any vessel to commit theft or any other crime with the intent or capability to use force in furtherance of that act."[9] The definition is wider than that adopted by the United Nations, which is restricted to acts that take place on the high seas. This is problematic because most acts of maritime violence take place in territorial waters or against ships already at port. The definition also excludes acts of violence committed at sea for purely political purposes. Such maritime "terrorism" is treated as a separate issue under international law[10] and has, in any event, been uncommon largely because land targets are more accessible.[11]

The International Maritime Bureau (IMB) has defined three types of piracy that currently take place in Southeast Asian waters:

- *Low Level Armed Robbery (LLAR).* Opportunist attacks mounted close to land
- *Medium Level Armed Assault and Robbery (MLAAR).* Violent attacks of theft involving serious injury or murder by well organized gangs, who are usually heavily armed and working from a "mother" ship
- *Major Criminal Hijack (MCHJ).* Well resourced and planned international criminal activity, using large gangs of highly trained and heavily armed operatives.[12]

At the low end of the spectrum are LLAR attacks. These assaults are typically directed against ships anchored at harbor and usually target cash or portable hi-value items, with an average theft of between $5,000 and $15,000 U.S. Perpetrators are generally armed with rudimentary weapons such as knives, pistols, and machetes, and tend to operate from small, maneuverable dinghies or other similar light craft. To a large extent, this type of piracy reflects the relatively lax security procedures that are adopted at many of the region's small- and medium-sized ports. More common and problematic are MLAAR attacks that are directed against ships traveling on the high seas or in territorial waters. This is a more serious form of piracy and one that, if carried out in narrow sea-lanes, has the potential to severely disrupt maritime navigation, particularly in cases where vessels are out of control because the crew has been detained or locked up. The scale of violence involved in these types of attacks is also of a greater magnitude than that for LLAR assaults given the higher probability of confronting crew members and the more powerful weapons used.[13]

At the high end of the spectrum are MCHJ assaults, involving the outright theft of ships and their subsequent re-conversion for the purposes of illegal trading (although, on a couple of occasions, ship owners have also arranged such attacks in order to defraud hull insurers). Known as the "phantom ship" phenomenon, this form of piracy follows a typical pattern. Vessels are first seized with their cargoes off-loaded into lighters at sea (such merchandise either being kept by the pirates themselves or sold off to private bidders). The ships are then fraudulently re-registered–generally under Panamanian, Liberian, Belize, St. Vincent, or Honduran flags–and issued with false documents to enable them to take on board a fresh payload. The new cargo is never delivered to its intended destination but is transferred to another vessel and taken to an alternate port where its is sold to a pre-arranged buyer who more often than not is a willing participant in the whole venture.[14]

According to the Bureau, syndicates can earn as much as $50 million U.S. from MCHJ attacks, with most groups showing very little hesitation to use lethal force if challenged. Most of the organizations involved in phantom ship attacks are either Chinese or of Chinese origin, belonging to either Cantonese, Shanghai, or Fuken gangs. These groups rely on intelligence gathered from operators and brokers–some of whom allegedly have direct access to government officials–located in China, Hong Kong, Taiwan, Singapore, Thailand, Malaysia, the Philippines, Indonesia, and Korea. This data provides the syndicates with accurate, up-to-date information about all relevant activities taking place in the shipping market such as contracts, cargoes, ships, routes, and buyers. It also ensures that they are kept abreast of any action that industry or law enforcement authorities intend to take to counter their activities.[15]

THE INCIDENCE OF PIRACY IN SOUTHEAST ASIA

In terms of statistical incidence, the threat posed by piracy in Southeast Asian waters has exhibited a rising trend since the end of the Cold War. As Table 10.1 illustrates, actual and attempted attacks rose from 3 in 1989, to

60 in 1990, before reaching an all time high of 160 in 1999. Overall, 989 separate attacks were recorded between 1989 and 1999, representing an annual average incident rate of 90.[16] Since 1991, these assaults have been of sufficient quantity to consistently designate Southeast Asia as the most pirate-prone region of the world.

The actual problem of piracy in Southeast Asian waters is undoubtedly far more serious than these figures suggest. According to Eric Ellen, the former Executive Director of the International Maritime Bureau (IMB) in London, in most cases law enforcement authorities and data collecting bodies only register incidents that occur on the high seas with perhaps less than half of these attacks actually reported in the first place. Ellen maintains that most ship owners are reluctant to report piracy attacks against their vessels for two main reasons. First, because of the costs (and delays) that subsequent legal investigations will necessarily entail, most of which will be borne by the companies themselves. Second, due to the fear that reporting attacks will serve to merely increase maritime insurance premiums.[17] Indeed, it has been alleged by the RPC in Kuala Lumpur that it is actually written into several

Table 10.1

Actual and Attempted (Reported) Piracy Incidents in Southeast Asia

Year	***Number of Reported Attacks***	***Percentage of Global total***
1989	03	N/A
1990	60	N/A
1991	102	95%
1992	69	65%
1993	75	73%
1994	62	69%
1995	116	62%
1996	139	61%
1997	108	44%
1998	95	50%
1999	160	56%

N/A means nonapplicable.

Source: ICC-IMB Regional Piracy Center, Kuala Lumpur.

shipping associations' manuals of procedures that pirate attacks should specifically not be reported to any law enforcement or information gathering body, official or not.[18]

Several reasons account for the high incidence of piracy in Southeast Asia. First has been the massive increase in commercial maritime traffic in the region, which has provided a ready supply of potential targets. Due to advances in technology, many of these vessels are now able to operate with little more than a skeleton crew, making the pirates' task of gaining control of the ship that much easier. Second, the post-Cold War "peace dividend" between Washington and Moscow has led to a greatly reduced superpower naval presence in Southeast Asia, something that has greatly diminished the deterrent effect on pirates operating in the region. Third, many of the coastlines in Southeast Asia pose enormous problems in terms of effective surveillance and monitoring due to their expansive and diffuse character. Difficulties in this regard have been especially acute in the waters around the Philippines and Indonesia—currently the two most "pirate prone" states in the region.

Fourth, it is almost certain that several gangs have benefited from the provision of logistical, intelligence, and financial support provided by corrupt officials on the Indonesian and southern Chinese mainlands. This backing has proved crucial in allowing pirates to pinpoint potential targets, quickly offload stolen cargo, escape arrest, and generally operate on a more sophisticated level.[19] Finally, the fall out of the 1997–1998 Asian economic collapse had a marked impact on piracy and other forms of organized illicit activity in Southeast Asia. In particular, it encouraged people to become involved in maritime (and other) crime as a way of "compensating" for falling wages, higher prices, and joblessness while, simultaneously reducing the availability of resources the littoral states were able to devote to anti-piracy patrols and coastal surveillance.[20]

THE SPECIFIC GEOGRAPHIC DISTRIBUTION OF PIRACY ATTACKS IN SOUTHEAST ASIA

Although Southeast Asia, as a whole, is the most pirate-prone region of the world, the actual distribution of attacks in this part of the world has shown some interesting changes during the 1990s, the specifics of which should be noted. Between 1990 and 1992, the IMB identified the waters between the Malacca and Singapore Straits as the most dangerous for piracy attacks both regionally and internationally, with assaults in this particular vicinity accounting for roughly half of all piracy attacks that took place in Southeast Asia during the two year period.[21] The high incidence of piracy along this maritime corridor essentially resulted from operational limitations that serve to increase the vulnerability of ships. The seaway is extremely crowded (up to

2,000 ships a day pass through the Malacca Straits from the South China Sea to the Indian Ocean), subject to the formation of sudden sandbanks/rain squalls, beset by numerous shallow points and is only eleven nautical miles wide at its southern end.[22] These factors necessarily force ships to reduce speed and follow a traffic separation system to ensure safe passage and ease congestion, making them especially susceptible to opportunist attacks mounted from Bantam Island and the neighboring spread of 180 islands in the Phillip Channel.[23]

Between 1993 and 1995, there was a significant shift in the focus of piracy in Southeast Asia, away from the Malacca and Singapore Straits and toward two distinct regions in the South China Sea:

- The territorial waters around Hong Kong and Macau
- An extended zone bounded by Hong Kong, Luzon (the Philippines), and Hainan Island (China)—the so-called HLH "terror triangle"

A total of 122 separate attacks were recorded in these areas during this period, representing 51.7 percent of all the assaults that took place in Southeast Asia at this time.[24] Two main reasons have been put forward to explain this change in regional locus:

- The initiation in 1992 of effective littoral antipiracy patrols by Malaysia, Indonesia, and Singapore, accounting for the drop in Malacca and Singapore Straits attacks
- Increased activity throughout the South China Sea by Hong Kong- and Macau-based groups, allegedly with the backing of corrupt elements of the southern Chinese bureaucracy[25]

Since 1995, however, there has been a dramatic drop in the number of incidents reported in these regions, with the RPC's database recording only 37 actual/attempted attacks between 1996 and 1999.[26] Reasons for the decline vary. The IMB maintains that it is simply due to a lack of reporting from the ports in China, Hong Kong, and Macau, and that gangs continue to operate more or less freely in this part of the region.[27] Others such as the Malaysian Institute of Maritime Affairs (MIMA) and the Thai Institute of Advanced Naval Studies (TIANS) believe attacks *have* declined, attributing this to a crackdown on corrupt officials in the southern provinces by the central government. Such action, according to MIMA and TIANS officials, was precipitated to reaffirm Beijing's maritime trading credentials, the reputation of which had begun to suffer badly as a result of repeated allegations concerning Chinese complicity in regional pirate attacks.[28]

Whatever the truth of the matter, it is the seas off the Philippines and especially Indonesia that have been most "pirate prone" in recent years, jointly accounting for 72 percent of all actual/attempted attacks recorded in Southeast Asia between 1996 and 1999.[29] Explanations for the high incidence of piracy in these two areas essentially relate to the extreme difficulty of effectively monitoring the extremely large and diverse Philippine and Indonesian coastlines,[30] combined with poor port security.

Further fueling the incidence of attacks has been a growing incentive for maritime crime since 1996. In Indonesia, this latter consideration stems from the fall-out of the Asian financial collapse between 1997–1998. The plummeting value of the rupiah since 1997 resulted in widespread social violence and spiraling crime as individuals struggle to come to grips with falling wages and rampant inflation. Rising illegality on land has been mirrored by growing lawlessness on the country's territorial seas, something that has been further encouraged by the lack of resources Jakarta is currently able (or prepared) to devote to the maritime realm.[31] During 1999, for instance, 113 actual or attempted attacks took place on Indonesian waters, which is 71 percent of the annual Southeast Asian total and just under double the number of incidents recorded in 1998. Thirteen attacks were also recorded in the Indonesian part of the Singapore Straits (compared to one in the previous three years), a resurgence of activity that again reflects Jakarta's growing inability to effectively discharge its maritime monitoring and enforcement responsibilities.[32]

In the Philippines, incentives for maritime crime have been heightened by the growing marginalization and fragmentation of two radical separatist movements in the south of the country, the Moro Islamic Liberation Front (MILF) and Abu Sayyaf Group (ASG). Both organizations rejected the 1996 Davao Consensus[33] signed between then President Fidel Ramos and Nur Misauri, the leader of the mainstream Moro National Liberation Front (MNLF), and have since progressively degenerated into loose collections of "lost commands" motivated less by ideology and more by greed.[34] The rising number of kidnappings and extortion on Mindanao reflect this transition, as does the increased prevalence of attacks that have been instigated against merchant vessels in the Moro Gulf and Sulu Sea areas since 1996.[35] Overall, approximately 16 percent of all attacks that took place in the South China Sea during the 1996–1999 period occurred in Philippine waters, a quarter and half of which were connected to the profiteering activities of renegade factions within the MILF and the ASG.[36]

IMPACT AND THREAT TYPOLOGIES

Contemporary piracy has significant implications for regional stability in Southeast Asia for at least four reasons. First, and at the most basic level, it constitutes a direct threat to the lives and welfare of the citizens of a variety of flag states. According to the IMB, most cases of piracy reported to the RPC now involve casualties of one sort or another. As Figure 10.1 portrays, the extent and seriousness of this violence has steadily increased during the 1990s.[37] During 1998 and 1999 alone, the Bureau documented a total of

965 aggressive acts in Southeast Asia, including 740 cases of hostage taking and intimidation, 81 assaults, 61 injuries, and most notably, 80 deaths.[38] As NUMAST, the maritime trade union and professional organization, observes: "The necessities of normal diplomacy should not obscure the fact that British nationals are being threatened with extreme violence. The present intolerable situation should be approached by the U.K. Government just as firmly as if British tourists were being attacked whilst [taking a holiday] in a [foreign] country."[39]

Quite apart from the risk of physical death and injury, many seafarers that have been subjected to a pirate attack have suffered considerable mental trauma. Many do not fully recover from the experience, and even though the chances of a repeat attack may be slight, they never go to sea again.[40] Despite this, the human cost involved in pirate attacks is something that rarely receives the attention it deserves, largely because assaults are directed against less-than-visible targets. Again NUMAST: "If you had civilian aircraft being threatened or bazookas being fired at train drivers there would be a public outcry. Because it's shipping, it's out of sight, out of mind and nothing is done."[41]

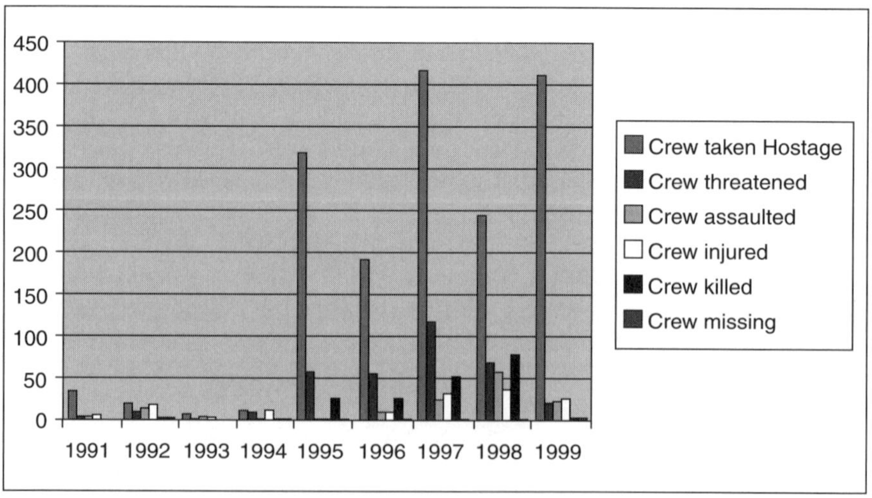

Figure 10.1

Types of Violence Committed to Ship Crew, 1991–1999.

Source: ICC-IMB Regional Piracy Center, Kuala Lumpur.

Second, piracy has a direct economic impact in terms of fraud, stolen cargoes, and delayed trips and could potentially undermine a maritime state's trading ability. As noted previously, ship owners are often required to pay their own legal expenses in terms of post-attack investigations, and they always have to bear the costs of canceled/interrupted onward journeys. Indeed, the IMB maintains this is one of the main reasons why attacks tend not to be reported to law enforcement authorities in the first place. The costs of MCHJ attacks can be particularly exorbitant, and on a few occasions, consignees have had to pick up the entire loss from phantom ship frauds.[42] Overall, it is estimated that at least $1 billion U.S. is lost to piracy every year.[43]

A reputation for vulnerability to piracy will also do little for the international standing of a trading country and could lead to a boycott of its port facilities. Between 1992 and 1995, for instance, Hong Kong gained a reputation as a port of "extreme danger" with both NUMAST and the Japan Shipowner's Association threatening to reduce and redirect trade until maritime surveillance and security was substantially increased. If carried out, threats of this type obviously have the potential to affect seriously any maritime state or territory that relies on having world-class and safe port facilities in terms of income generation.[44]

Third, piracy can play a pivotal role in undermining and weakening political stability by encouraging corruption among elected government officials. Particular attention has been leveled against China in this regard, with numerous shipping associations and maritime bodies decrying the complicity of southern officials in arranging or otherwise facilitating both low- and high-end attacks.[45] Victims of pirate attacks in Jakarta's territorial waters have also made similar allegations concerning corrupt Indonesian officials. In one notable testimony given before the RPC, a Master reported that within minutes of foiling an attack on his vessel by eight "professional looking men, a fast Indonesian gunboat went past our ship toward the pirates. It is my opinion—and that of our Singapore agents—that these two groups work together."[46]

Finally acts of piracy have the potential to cause major environmental and pollution damage, particularly if they take place in crowded sea-lanes traversed by heavily laden petroleum tankers. In many instances, pirates will either tie up or throw overboard crew members after raiding their vessels. Known as "rogue ships," these uncontrolled craft obviously have enormous implications with respect to the safety of maritime navigation, and they greatly increase the possibility of a highly damaging mid-sea collision. Analysts have posited several worrying scenarios that could result from accidents of this sort. One of the more serious envisages a crash involving a liquefied natural gas (LNG) tanker. The bulk of the world's LNG trade passes through the South China Sea, with Indonesia, in particular, playing a major role in terms of regional supply.[47] The growing lawlessness around the archipelago's territorial waters has greatly exacerbated the danger of an uncontrolled "pirated" ship crashing into one of these "floating bombs"—the effects of which could have untold implications for surrounding fishing grounds and fragile maritime ecosystems.

Possibly an even more serious scenario would be one involving a heavily laden oil tanker. The resulting discharge of petroleum would not only cause irreparable damage to maritime life and other off-shore resources; if left to drift, it would also seriously degrade large tracts of fertile coastal low land.[48]

In March 1989, the world was appalled at the ecological and environmental destruction caused by the Exxon Valdez disaster in Alaska. Although not caused by piracy, the catastrophe graphically illustrated what can happen when a major oil discharge occurs. Conservative figures have since estimated that up to 1800 miles of coastline were affected, some of which may never fully recover.[49] Luckily much of this area is sparsely populated. However, transpose the Exxon Valdez experience to a similar incident occurring in the extremely congested Southeast Asian Phillip Channel as a result of a deliberate pirate attack, and the potential for an environmental disaster of the highest order becomes manifest.[50]

THE ROLE OF SATELLITE AND INFORMATION TECHNOLOGY IN COUNTER-PIRACY

The growing frequency and seriousness of maritime violence and crime in Southeast Asia has prompted moves on the part of the international shipping industry to look at innovative ways of better protecting ships sailing in pirate prone areas. Apart from issuing guidelines on practical measures that crews should take to prevent piracy attacks, considerable attention has focused on more "high-tech" solutions, particularly those that incorporate satellite-based positioning and tracking systems. A number of initiatives have already been developed, three of which have particular relevance with respect to counter-piracy.

The first is a system known as Time Division Multiple Access (TDMA). Developed by Racal Tracs, a world leader in maritime positioning, this is an intelligent radio data network that integrates advanced UHF and VHF communications with Global Positioning Satellite (GPS) accuracy. TDMA allows continuous real-time positional messaging and is capable of handling up to 250 mobile reports per second. Apart from providing continuous tracking information from a wide variety of applications, the system can also be used to broadcast data such as vessel status, thus permitting reliable signal transmission in "difficult" conditions without operator intervention.[51]

The second is a satellite-based network that enables a centralized control center to track and automatically communicate with an unlimited number of vessels or sensors anywhere in the world. Known as SAT/C and again developed by Racal Tracs, this network operates through Inmarasat satellites, and it provides automatic GPS position and status reports in addition to offering two-way messaging facilities based on fax, e-mail, telex, or X25. Vessel reports can be manually initiated by the control center or automatically triggered by events such as a significant reduction in a ship's speed or a dramatic departure from a pre-defined course, which would strongly indicate that a hijack had taken place.[52]

A third, cheaper solution that is currently being promoted by the IMB is ShipLoc, a basic maritime satellite tracking device produced as a result of a joint collaborative venture between the Bureau and *Collecte Localisation Satellites* (CLS).[53] This system involves a tiny transmitter. It can be powered either by the onboard power supply or a back-up battery that beams a ship's position to a specially dedicated IMB Web page up to 15 times a day. If a vessel fails to radio in its daily report, owners merely have to log onto the Internet to ascertain where their vessels is and to determine whether a major departure in scheduled course has taken place. The device can be rented for as little as $280 U.S. a month and is now commercially available for general distribution.[54]

All of these initiatives have been recognized as valuable developments in the regional fight against piracy. They can all be covertly installed on vessels to provide position reports without the knowledge of the crew and, at least in the case of TDMA and SAT/C, can be configured in such a way that control centers are immediately alerted if antennas or power supplies are disconnected. Tracking of this sort will make it far more difficult for pirates to seize a ship (or its cargo) and return it to commercial service under a new name and paint scheme. Moreover, the systems exhibit a valuable dual-use quality, serving as a useful management tool for ship operators when not required for specific antipiracy duties.

In addition to tracking, aerial imagery has also been suggested as a logical way to enhance the effectiveness of coastal surveillance, particularly for states that have expansive archipelagic shorelines (such as Indonesia and the Philippines). One idea has been to make E-3 Sentry AWACS aircraft available for regional antipiracy patrols. While these planes are able to detect maritime traffic of a ship size, they not well suited for discriminating between vessels and have only marginal utility in terms of monitoring small seaborne craft. A more realistic option suggested by William Carpenter and David Wiencek is to use inverse synthetic aperture radar (ISAR) or low light level television (LLTV) of the sort installed on U.S. Navy P-3 aircraft. This type of aerial imagery has a far greater capability to distinguish between ships and could conceivably be used to differentiate pirate craft from other legitimate vessels such as fishing trawlers.[55]

Japan is already making moves in this direction. Piracy has long occupied the minds of policy makers in Tokyo, given the country's reliance on strategic sea lines of communication (SLOCs) that pass directly through the South China Sea. Indeed over the past couple of years, Japan has increasingly voiced support for its Self Defense Forces (SDF) to play a more active role in antipirate initiatives and patrols, especially in regard to helping augment unilateral and collaborative Southeast Asian endeavors.[56] In this context, the government recently announced plans to request funds for the purchase of two long range reconnaissance aircraft that could be made available for antipirate surveillance missions along the Chinese, Taiwanese, Vietnamese, and Indonesian coasts. The planes, which would not require basing in the region (having the ability to fly from Japan to the Strait of Malacca and back), would be equipped with state-of-the-art surveillance and electronics packages and used to provide a long-reach component to the Coast Guard's existing airborne monitoring capacity.[57]

A logical extension of aerial coastal surveillance is the use of commercial satellite imagery in antipiracy missions. Certainly space-based observational systems could play a role in facilitating post-attack investigations, especially in the context of helping to identify suspected phantom vessels at specific ports of call and anchorage. Although definitive identifications are probably not yet possible, high-resolution commercial satellite technology of the sort currently available (presently down to one meter) does appear to have the ability to pick up distinct ship signatures. Such information could be used to provide at least partial evidence of vessel identity that could then be taken by owners to governments, international maritime organizations, and insurance companies. Perhaps the clearest indication of this potential is the extensive use that has already been made of Radarsat and ERS imagery to monitor and trace the source of ship-based pollution, such as sewage and oil.[58]

High-tech solutions are also being sought to augment harbor and anchorage security. LLAR attacks have been virtually eliminated in the Port of Singapore, for instance, as a direct result of the installation of Close Circuit Television (CCTV). Each terminal has its own centralized CCTV control room, which is monitored on a twenty-four–hour basis by a specially designated Auxiliary Port Police Force.[59] The IMB has urged other states in the region to install similar devices and encouraged further research and development into electronic threat-mitigation devices. A boat security service recently developed by Racal in conjunction with MarineGuard, a leading maritime security company, could have some application in this regard. The system combines GPS positioning with a global system for mobile (GSM) communications into an integrated network that can automatically report if a boat at anchor has been broken into, or even if there has been any change in its bilge water level. It is anticipated that the new service will be able to provide ship owners with a continuous watch on their vessels within over 80 countries worldwide.[60]

CONCLUSIONS

There can be little doubt that piracy has emerged as a growing and significant threat to maritime security in Southeast Asia. Attacks not only have implications for the safety of shipping in one of the most important maritime corridors in the world; they also have the potential to trigger major environmental disasters. All present indications suggest that the level of piracy in Southeast Asia will continue to grow, fueled by a ready supply of targets as well as increased motivations and opportunities for engaging in armed maritime crime.

Satellite and information technology offers a valuable and, as yet, largely unused modality for enhancing the regional fight against piracy. In particular, it offers an innovative way to enhance transparency and general security awareness in what has traditionally been a highly opaque operating environment. Incorporating these types of contingencies does not necessarily have to entail a large capital outlay, and certainly systems such as ShipLoc should fall within the budgetary parameters of most maritime owner-operators.

Facilitating the provision of more effective aerial and satellite imagery to augment coastal surveillance would also seem to be an especially appropriate way for extra-regional flag states to contribute to antipiracy measures in Southeast Asia. This type of assistance is not only more politically acceptable (and realistic) than sending armed escorts to accompany merchant vessels operating in high-risk areas—something vigorously advocated by ship masters in Southeast Asia—it would additionally serve as a useful way to enhance general confidence building in the region, which could, in turn, help to mitigate other sources of maritime tension and instability.[61]

It is important to remember that security environments, both in terms of threat and response, are neither static nor "zero sum" in nature. This is as true for the sea/piracy realm as it is for more conventional land-based contexts. Certainly the maritime marauders of the distant and even recent past bear little, if any, relevance to the organized, well-financed, and ruthless syndicates that now threaten global shipping commerce. By the same token, there is now a far greater array of available technologies that can be brought to bear against this particular menace. The crucial question is whether southeast Asian politicos, shipping associations, and international flag states will show the necessary foresight to adapt to the challenges posed by modern-day piracy and take advantage of these new modalities.

NOTES

1. For the purposes of this paper, Southeast Asia is taken to include the ASEAN states (Singapore, Thailand, Malaysia, the Philippines, Indonesia, Brunei, Vietnam, Laos, Cambodia, and Burma) in addition to southern China.
2. F. Moorhead, *A History of Malaya and Her Neighbors* (Kuala Lumpur: Longmans of Malaysia, 1957), 134. See also H. Giles, *The Travels of Fa-Hsien, 399–414 AD* (London, 1956).
3. L. Wright, "Piracy in the Southeast Asian Archipelago," *Journal of Oriental Studies* 14 (1976), 26.

4. Nicholas Tarling, *Piracy and Politics in the Malayan World: A Study of British Imperialism in Nineteenth Century Southeast Asia* (Melbourne: F.W. Cheshire, 1963), 6–8.
5. Robert Beckman, Carl Grundy-Warr, and Vivian Forbes, "Acts of Piracy in the Malacca and Singapore Straits," *International Boundaries Research Unit (IBRU) Maritime Briefing* 1:4 (1994), 4.
6. Michael Pugh, "Piracy and Armed Robbery at Sea: Problems and Remedies," *Low Intensity Conflict and Law Enforcement* 2:1 (1993), 2.
7. Jack Roberts, "Piracy in the Far East," *INTERSEC* 1:5 (1991), 160. For a detailed account of piracy in the Gulf of Thailand during this period see Pascal Boulanger, "The Gulf of Thailand," in Eric Ellen, ed., *Piracy at Sea* (Paris, France: International Chamber of Commerce [ICC] Publishing SA, 1989), 83–97.
8. See International Maritime Bureau (IMB), *Piracy and Armed Robbery Against Ships. Annual Report, 1999* (London: ICC, January 2000), 3.
9. This is the definition used by the ICC's IMB.
10. Political violence at sea is covered by the 1988 Rome Convention for the Suppression of Unlawful Acts Against the Safety of Maritime Navigation.
11. For further details see William M. Carpenter and David G. Wiencek, "Maritime Piracy in Asia," in William M. Carpenter and David G. Wiencek, eds., *Asian Security Handbook. An Assessment of Political-Security Issues in the Asia-Pacific Region* (New York: M.E. Sharpe, 1996), 81–2; Jon Vagg, "Rough Seas? Contemporary Piracy in Southeast Asia," *British Journal of Criminology* 35:1 (Winter 1995), 64; B. Parritt, *Security at Sea: Terrorism, Piracy and Drugs. A Practical Guide* (London: The Nautical Institute, 1991), 13–17; and S. Menefee, "Piracy, Terrorism and the Insurgent Passenger: A Historical and Legal Perspective," in N. Ronzitti, ed., *Maritime Terrorism and International Law* (Dordrecht: Martinus Nijhoff, 1990), 56–61.
12. See Edward Fursdon, "Sea Piracy—or Maritime Mugging?" *INTERSEC* 5:5 (May 1995), 166.
13. Edward Fursdon, "Sea Piracy—or Maritime Mugging?," *INTERSEC* 5:5 (May 1995), 166; Stanley Weeks, "Law and Order at Sea: Pacific Cooperation in Dealing with Piracy, Drugs and Illegal Immigration," in Sam Bateman and Stephen Bates, eds., *Calming the Waters. Initiatives for Asia-Pacific Maritime Cooperation* (Canberra: The Strategic and Defense Studies Centre, 1996), 44.

14. See Robert Redmond, "Phantom Ships in the Far East," *INTERSEC* 6:5 (May 1996), 168–169; Michael Pugh, "Piracy and Armed Robbery at Sea," *Low Intensity Conflict and Law Enforcement* 2:1 (1993), 9; and "Phantom Vessels the Latest Tactic in Asian Piracy," *South China Morning Post* (July 21, 1994).
15. Personal correspondence between author and Eric Ellen, IMB, London, January 1997. See also "Dead Men Tell No Tales," *The Economist* (December 18, 1999). See also Edward Fursdon, "Sea Piracy–or Maritime Mugging?," *INTERSEC* 5:5 (May 1995), 171; Robert Redmond, "Phantom Ships in the Far East," *INTERSEC* 6:5 (1996), 168; "The Virtual Immunity Enjoyed by Pirates," *Lloyd's Shipping Economist* (November 1996), 6–7; "Shipowners Deny Piracy is a Problem," *Commercial Crime International* 14:8 (January 1997), 1; "Unfurl the Skull-and-Crossbones?" *Foreign Report* (July 11, 1996); "Phantom Vessels the Latest Tactic in Asian Piracy," *South China Morning Post* (July 21, 1994); "Pirates Employ New Tactics in Asian Waters," *New Straits Times* (September 1, 1994); and "Tracking Down the 'Phantom Ship,'" *New Straits Times* (October 17, 1994).
16. IMB, *Piracy and Armed Robbery Against Ships. Annual Report,* 1999, (London: ICC, January 2000), 3.
17. Personal correspondence between author and Eric Ellen, IMB, London, January 1997. See also "Piracy Becomes World Threat," *Sunday Telegraph* (May 25, 1997); and "Piracy Reports Seen as a Waste of Time," *South China Morning Post* (May 12, 1993).
18. Personal correspondence between author and John Martin, Regional Piracy Centre (RPC), Kuala Lumpur, April 1997.
19. Personal correspondence between author and IMB officials, London and Kuala Lumpur, 1997–1999. See also IMB, *Piracy and Armed Robbery Against Ships. A Special Report* (London: ICC, 1997), 4–5.
20. Personal correspondence between author and Asian analysts with Control Risks Group (CRG), London, February 1998.
21. IMB, *Piracy Annual Report 1992* (London: ICC, 1993), 4. See also Amitav Acharya, *A New Regional Order in Southeast Asia: ASEAN in the Post-Cold War Era*, Adelphi Paper 279 (London: International Institute of Strategic Studies, 1993), 37; Stanley Weeks, "Law and Order at Sea: Pacific Cooperation in Dealing with Piracy, Drugs and Illegal Immigration," in Sam Bateman and Stephen Bates, eds., *Calming the Waters. Initiatives for Asia-Pacific Maritime Cooperation* (Canberra: The Strategic and Defense Studies Centre, 1996), 46; and Robert Beckman, Carl Grundy-Warr, and Vivian Forbes, "Acts of Piracy in the Malacca and Singapore Straits," *International Boundaries Research Unit (IBRU) Maritime Briefing* 1:4 (1994), 13–14.

22. G. Naidu, "The Strait of Malacca and the Straits States: A Malaysian Perspective," paper presented before the International Conference on the Strait of Malacca: Meeting the Challenges of the 21st Century (Kuala Lumpur, June 14–15, 1994, 11–12.
23. See Amitav Acharya, *A New Regional Order in Southeast Asia: ASEAN in the Post Cold War Era*, Adelphi Paper 279 (London: International Institute of Strategic Studies, 1993), 37-8; Robert Beckman, Carl Grundy-Warr, and Vivian Forbes, "Acts of Piracy in the Malacca and Singapore Straits," *International Boundaries Research Unit (IBRU) Maritime Briefing* 1:4 (1994), 14; Neil Renwick and Jason Abbott, "Piratical Violence and Maritime Security in Southeast Asia," *Security Dialogue* 30:2 (1999), 185.
24. IMB, *Piracy and Armed Robbery Against Ships. Annual Report 1996* (London: ICC, 1997), 3. See also "Piracy Hotspot Shifts to Triangle of China, Philippines and Hong Kong," *Shipping Times* (May 25, 1993); "HK-Luzon-Hainan Triangle Still Prone to Pirate Attacks," *Business Times* (May 18, 1995); and "Piracy on the Rise in HK-Luzon-Hainan Area," *New Straits Times* (May 22, 1995).
25. Personal correspondence between author and IMB officials, London and Kuala Lumpur, 1997–1999. See also Stanley Weeks, "Law and Order at Sea: Pacific Cooperation in Dealing with Piracy, Drugs and Illegal Immigration," in Sam Bateman and Stephen Bates, eds., *Calming the Waters. Initiatives for Asia-Pacific Maritime Cooperation* (Canberra: The Strategic and Defense Studies Centre, 1996), 46; Robert Beckman, Carl Grundy-Warr, and Vivian Forbes, "Acts of Piracy in the Malacca and Singapore Straits," *International Boundaries Research Unit (IBRU) Maritime Briefing* 1:4 (1994), 13–20; "Unfurl the Skull and Crossbones?" *Foreign Report* (July 11, 1996); and Neil Renwick and Jason Abbott, "Piratical Violence and Maritime Security in Southeast Asia," *Security Dialogue* 30:2 (1999), 187.
26. IMB, *Piracy and Armed Robbery Against Ships. Annual Report*, 1999, (London: ICC, January 2000), 3; IMB, *Piracy and Armed Robbery Against Ships. Report for the Period January 1–September 30, 1999* (London: ICC, July 1999), 5.
27. Personal correspondence between author and Eric Ellen, IMB, London, February and November 1998.
28. Personal correspondence between author and MIMA/TIANS representatives, Kuala Lumpur and Bangkok, April–July 1997. See also "China to Impose Strict Guide to Help Quell Piracy Attacks," *Daily Commercial News*, (March 22, 1994); and "China Pins Piracy on 'Rebel' Lawmen," *The South China Morning Post* (May 14, 1994).

29. IMB, *Piracy and Armed Robbery Against Ships. Annual Report,* 1999, (London: ICC, January 2000), 3; IMB, *Piracy and Armed Robbery Against Ships. Report for the Period January 1– September 30, 1999,* (London: ICC, July 1999) 3.
30. According to some experts, even technically advanced states such as the United States would have difficulty mounting effective surveillance over coastlines as vast and diverse as those of Indonesia and the Philippines.
31. Personal correspondence between author and Asian analysts with CRG, London, February 1998.
32. IMB, *Piracy and Armed Robbery Against Ships. Annual Report,* 1999, (London: ICC, January 2000), 3.
33. For an overview of the Davao Consensus and the Islamic separatist insurgency in Mindanao see Peter Chalk, "The Davao Consensus: A Panacea for the Muslim Insurgency in Mindanao," *Terrorism and Political Violence* 9:2 (1997), 79–98.
34. Personal correspondence between author and the Philippine National Police Directorate for Intelligence, Manila, June 1998.
35. Personal correspondence between author and Jayant Abyankar, IMB, November 1998. Similar comments were expressed to the author by Ms. Emma Sarne, Senior Advisor to the Philippine National Security Council during the combined meetings of the Council for Security Cooperation in Asia Pacific (CSCAP) Transnational Crime and Maritime Cooperation Working Groups (Wollongong, New South Wales, Australia, November 1999). See also Anthony Forster, "An Emerging Threat Shapes Up as Terrorists Take to the High Seas," *Jane's Intelligence* Review, 10 (July 1998) 43–45; Neil Renwick and Jason Abbott, "Piratical Violence and Maritime Security in Southeast Asia," *Security Dialogue* 30:2 (1999), 187-88.
36. IMB, *Piracy and Armed Robbery Against Ships. Annual Report, 1999,* (London: ICC, January 2000), p. 3. See also the narration of attacks outlined in the IMB annual piracy reports, 1996-1999. It needs to be acknowledged that a drop occurred in Philippine activity during 1999, with only six attacks recorded by the RPC during the year. While Manila maintains that this is the result of more effective coastal patrols that have been initiated in the largely lawless Moro Gulf and Sulu Sea areas, it is just as likely a function of under or misreporting, something that is rife in this part of the Philippines.
37. Personal correspondence between author and Jayant Abyankar, IMB, London, November 1998.

38. IMB, *Piracy and Armed Robbery Against Ships. Annual Report, 1999*, (London: ICC, January 2000), 7. See also "The Spread of Piracy," *Foreign Report* no. 2576 (January 2000), 1. It should be noted that most of the fatalities occurred in 1998, with only three deaths reported in 1999. However, the decline in numbers stems more from luck–many of the seamen thrown overboard during the year's attacks were subsequently rescued–than to any departure from violence on the part of the pirates. In many ways indicative of this, there were 408 cases of hostage-taking in 1999–the highest figure on record.
39. *NUMAST Telegraph* 25:7 (July 1992, piracy supplement), i.
40. Eric Ellen, "Piracy," *Shipping at Risk*, 49–50.
41. Captain Graeme Hicks, Secretary of NUMAST. Cited in "Those in Peril on the Sea," *The Economist* (August 9, 1997), 50.
42. Jayant Abyankar, "Phantom Ships," in Eric Ellen, ed., *Shipping at Risk* London: ICC Commercial Crime Services, 1997), 58–74.
43. "Modern Pirates Terrorize the Seas," *Associated Press* (November 6, 1999); "Beware the Jolly Roger," *Foreign Report,* no. 2535 (March 11, 1999), 1–2.
44. See, for instance, Robert Beckman, Carl Grundy-Warr, and Vivian Forbes, "Acts of Piracy in the Malacca and Singapore Straits," *International Boundaries Research Unit (IBRU) Maritime Briefing* 1:4 (1994), 14; Kazuo Takita and Bob Couttie, "ASEAN Pressured to Act against Pirates," *Lloyds List* 29 (May 1992), 3; Pugh "Piracy and Armed Robbery at Sea," *Low Intensity Conflict and Law Enforcement* 2:1 (1993), 11; Brian Walters, "Coastal Patrol Boats," *The Asian Defense Journal* 5 (1993), 56; and "Port's Reputation Attacked by Seamen," *South China Morning Post* (November 6, 1992).
45. Personal correspondence between author and IMB officials, London, February and November 1998. See also "Sailors Walk the Plank into Brutal New Era of Asian Piracy," *The Australian* (February 3, 1999); "High-Tech Pirates Ravage Asian Seas," *Washington Post* (July 5, 1999); and "Piracy: The World's Oldest Profession," *The Scotsman Weekend Magazine* (August 16, 1997). Several IMB officials have gone further in their allegations of Chinese involvement in piracy, claiming that Beijing has been deliberately prepared to back "anti-smuggling" rings as a way of forcibly exerting its sovereignty throughout the South China Sea.
46. *NUMAST Telegraph* 25:7 (July 1992, piracy supplement), iii. See also Michael Richardson, "Crackdown on Piracy," *The Asia-Pacific Defense Reporter* (October–November 1992), 34; Neil Renwick and Jason Abbott, "Piratical Violence and Maritime Security in

Southeast Asia," *Security Dialogue* 30:2 (1999), 187; and "Sailors Walk the Plank into a Brutal New Era of Asian Piracy," *The Australian* (February 3, 1999).

47. Written correspondence between author and Erik Kreil, August 2000.

48. See, for instance, Amitav Acharya, *A New Regional Order in Southeast Asia: ASEAN in the Post-Cold War Era*, Adelphi Paper 279 (London: International Institute of Strategic Studies, 1993), 38; Michael Grey, "Piracy–a Hazard of our Times," *BIMCO Bulletin* 2 (March–April 1992), 23 "Piracy Becomes World Threat," *The Sunday Telegraph* (May 25, 1997); and "Terror on the High Seas," *The Australian* (November 21, 1995).

49. IMB, *Piracy and Armed Robbery Against Ships: A Special Report* (London: ICC Commercial Crime Services, 1997), 32.

50. The Malacca Strait has already witnessed some notable "near misses" in this regard. The most spectacular incident was a collision that occurred in September 1992 between the supertanker *Nagasaki Spirit* and a container ship *Ocean Blessing* at the northern end of the Malacca Strait. *Ocean Blessing*, which collided with the oil tanker, was apparently a "rogue ship," zigzagging across shipping lanes and possibly under the control of pirates. The collision resulted in a fire that killed all aboard *Ocean Blessing* and twenty members of *Nagasaki Spirit*. Although 13,000 tons of oil was spilled as a result of the incident, favorable weather conditions that day prevented its spread, with the majority quickly evaporating in the ensuing tanker blaze.

51. Racal Applications, Marine Transport Applications, Web site: http://www.racal-tracs.com/apps/martrans.htm.

52. Ibid.

53. *Collecte Localisation Satellites* is a leading international company that specializes in satellite-based location and data collection systems.

54. See IMB, *Piracy and Armed Robbery Against Ships. Annual Report, 1999*, (London: ICC, January 2000), 16 ; IMB, *Piracy and Armed Robbery Against Ships. Report for the Period January 1– March 31, 2000*, (London: ICC, July 1999), 10; and "Dead Men Tell No Tales," *The Economist* (December 18, 1999), 87–89.

55. William M. Carpenter and David G. Wiencek, eds., "Maritime Piracy in Asia," *Asian Security Handbook. An Assessment of Political-Security Issues in the Asia-Pacific Region* (New York: M.E. Sharpe, 1996), 86.

56. Comments made during the "Transnational Threats in Asia" International Conference, Hawaii, August 8–10, 2000.

57. "Bandits in the Global Shipping Lanes," *The New York Times Magazine* (August 21, 2000). It remains to be seen whether the Japanese Diet will endorse the requests for funds, which would, at least symbolically, represent a significant reformulation of the country's post-World War II pacifist constitution (and its emphasis on *territorial* self-defense forces only).
58. Communication with John Baker, RAND, August 2000.
59. Comments made Superintendent Bhopinder Singh, Intelligence Operations Division, during the Joint Meeting of the CSCAP Working Groups on Maritime Cooperation and Transnational Crime, Wollongong, Australia, November 1999.
60. Racal Applications, Marine Transport Applications, Web site: http://www.racal-tracs.com/apps/martrans.htm. An electronic tagging system, again developed by Racal, could also have some application in terms of augmenting port security. Based on a small-scale transponder that transmits electronic data to an interrogation antenna, it allows a central control to monitor the movement of both personnel and merchandise and could conceivably be used to provide automatic surveillance of cargo and important high-value items.
61. Much of this tension and instability derives from maritime territorial disputes, particularly claims relating to the following:
    - The Spratly Islands (variously contested by the Philippines, China, Taiwan, Vietnam, Brunei, and Malaysia)
    - The Paracel Islands (contested by China and Vietnam)
    - Natuna Island (contested by Indonesia and Vietnam)
    - Pedra Branca (contested by Malaysia and Singapore)
    - The islands of Sipadan, Sebatik, and Ligitan (contested by Malaysia and Indonesia).

    See Desmond Ball, "Arms and Affluence: Military Acquisitions in the Asia-Pacific Region," *International Security* (Winter 1993–1994), 88–89.

11

Conclusions

John C. Baker and David G. Wiencek

This book has examined the desirability and feasibility of creating a cooperative monitoring regime for the South China Sea from a variety of perspectives. Its main aim is to assess whether a new generation of commercial and civilian observation satellites can help contribute to diplomatic efforts to mitigate the risks of armed conflict that could arise from the long-standing territorial disputes over the Spratly Islands. The proposed cooperative monitoring regime involves taking a broader approach to the Spratlys problem by building on the common interests of the South China Sea littoral states in their cooperation on important problems, which include monitoring maritime environmental problems, responding to natural disasters, and perhaps even combating piracy. Attention is also given to the supportive role that interested governments and nonstate actors from beyond the region could play. We have addressed the following three key questions in exploring the prospects for a South China Sea cooperative monitoring regime:

- First, is there a compelling need for a cooperative monitoring regime to minimize the chances of an armed conflict in the Spratly Islands?
- Second, what are the key technical and operational factors that must be present to make cooperative monitoring feasible?
- Third, can a broader cooperative monitoring regime for the South China Sea be developed that addresses the common interests of the littoral states (both the claimants and nonclaimants to the Spratlys), as well as involves other concerned parties, including multinational and nongovernmental organizations (NGOs)?

THE NEED FOR COOPERATIVE MONITORING IN THE SOUTH CHINA SEA

The deceptively dangerous nature of the Spratly Islands disputes poses a challenge for formulating effective means to prevent an armed conflict from eventually occurring. A cyclical pattern has become too familiar whereby periods of increased tensions are followed by bilateral or multilateral rounds of diplomacy aimed at reaching temporary accommodations among the claimant states. Increased tensions are usually prompted by one or more of the Spratly claimants expanding or upgrading their permanent outposts. In comparison, a subsequent reduction in tensions often results from inconclusive diplomatic efforts to resolve, or at least diminish, the Spratly Islands disputes. The end result, however, is a persistent risk of conflict that includes the possibility of a serious armed conflict arising given the right (or wrong) confluence of foreign policy and domestic political developments. Added to this potentially worrisome combination is the unpredictability of inadvertent or unauthorized actions by military forces that are often in close contact and operating with less-than-perfect command, control, and communication links to their central authorities.

The unexpectedly bloody escalation of the protracted sovereignty disputes between Argentina and the United Kingdom over the Falklands (or Malvinas) in 1982 is a prudent reminder that political leaders, not to mention diplomats and military planners, can be lulled into a false sense of confidence that they can manage a persistent territorial dispute in order to prevent it from erupting into a limited or possibly more wide-ranging armed conflict. Similarly, the unexpected tensions arising over the collision between a Chinese fighter and an American reconnaissance aircraft flying in international waters off Hainan Island in April 2001 was a warning of how seemingly routine military activities can inadvertently become the catalyst for a rapid increase in the risk of armed conflict in the South China Sea region.

The complex sovereignty disagreements surrounding the Spratly Islands and their adjacent waters make it difficult—but not impossible—to imagine diplomatic and legal solutions for ending the territorial disputes. Efforts by the ASEAN countries to develop a "code of conduct" for diminishing the risk of conflict in the South China Sea have been complicated by China's insistence on dealing with the other claimants countries on a bilateral basis instead of through a multilateral forum, such as the ASEAN Regional Forum (ARF), or even through unofficial "track two" discussions. Nonetheless, some gradual progress is being made in engaging the claimant countries, including China, in a wider security dialogue that could lead to resolving the competing national claims of the Spratlys over the long run.

TECHNICAL AND OPERATIONAL FEASIBILITY OF COOPERATIVE MONITORING

Given the prospects for prolonged multinational competition over the Spratlys, this book has emphasized the need to consider a new tool now available to diplomats—new information technologies, specifically high-resolution commercial and civilian observation satellites, which could help reduce the risk of armed conflict from occurring. This book concludes that imaging satellites have some important advantages over the existing means of reconnaissance, such as low-level aerial flights and close-in naval patrols, that some states use today for monitoring activities in the South China Sea.[1]

First, commercial satellite images offer a much less intrusive means for monitoring what is occurring among the disputed Spratly Islands. The imagery analysis presented in this book provides a proof-of-concept that the new generation of commercial observation satellite imagery possess sufficient resolution to monitor politically sensitive developments, such as the substantial expansion of existing outposts or the construction of new ones. And unlike military aircraft and naval ships, space-based imaging sensors pose little risk of provoking a forcible or unauthorized reaction from the military forces stationed on a given island outpost that is being observed.

Second, commercial and civilian satellite imagery has the great advantage of being a credible data source that can be widely shared. Because third parties are the sources of the raw images, their overhead images will probably enjoy greater credibility and political acceptability when compared with the aerial photographs that are occasionally released by various states with claims to the Spratlys. Most importantly, these satellite imagery data are potentially available to a broad range of interested parties, including the South China Sea littoral states without claims to the Spratly Islands that do not regularly patrol these disputed waters with their air and naval forces, as well as multinational and nongovernmental organizations (NGOs). In this sense, the satellite imagery provides an unprecedented opportunity for "digital diplomacy" by providing both the Spratly claimants and the nonclaimants with a means for engaging in a dialogue over what is revealed in the satellite imagery. This process could have the beneficial effect of helping to provide a clear perception of the actions of the claimant states while providing a way for outside parties to express their concerns.

The imagery analysis presented in this book provides evidence that high-resolution satellite imagery is suitable for monitoring the Spratly Islands. The new generation of higher-resolution commercial and civilian observation satellites possess the resolution and frequency of coverage required to detect and identify politically provocative activities, such as establishing new outposts, that could occur among the disputed Spratly Islands. Our analysis also indicates that only a relatively small staff of experts would be required to perform the imagery interpretation and analysis functions necessary for supporting a cooperative monitoring regime focused on the Spratlys.

We also proposed taking a broad approach to the idea of developing a cooperative monitoring regime for the South China Sea to increase its political acceptability. Besides monitoring the disputed Spratlys, we recommend adopting a broader approach that takes advantage of the wide range of useful applications for satellite remote sensing data to foster patterns of cooperation involving all of the South China Sea littoral states. These applications include using satellite imagery for monitoring such common problems as maritime pollution, disaster warning and response, and perhaps even helping to combat the continuing problem of piracy.

Of course, satellite imagery alone cannot resolve the complex sovereignty issues underlying the Spratly Islands disputes. However, by shining a light on the activities of the disputants, we believe a cooperative monitoring regime using commercial and civilian satellite imagery might have a better chance of discouraging an armed conflict. Over the long-term, a cooperative monitoring regime could also become part of any eventual diplomatic settlement that is negotiated among the countries concerned with the Spratlys. Such a regime would offer a precedent for cooperation among the South China Sea littoral states, as well as provide a mechanism for monitoring their compliance with any legal obligations or unilateral pledges associated with resolving the Spratly Islands disputes.

THE POLITICAL CHALLENGE OF MOVING TO COOPERATIVE MONITORING

The prospects for developing a viable cooperative monitoring regime focused on the South China Sea ultimately depends on the participation of the various Spratly claimants, as well as other national governments and multinational organizations that are concerned with the potential security risks and economic opportunity costs of the unpredictable South China Sea dispute. Not surprisingly, the claimants to the Spratly Islands are bound to be ambivalent about participating in a cooperative monitoring regime. Each government will be very interested in receiving commercial satellite imagery data on the activities of other parties in the Spratly Islands area. However, they will be equally reluctant to have their own facilities and activities in the South China Sea subjected to the same scrutiny by other parties. Nonetheless, as more commercial and civilian observation satellites become operational, their neighbors, whether friends or enemies, will probably take advantage of this new source of satellite imagery for monitoring regional developments, including activities among the Spratly Islands. With the inevitable proliferation of high-resolution commercial satellite imagery, the key question is whether these countries will take the next step by employing this new information technology to engage in a cooperative monitoring regime for the South China Sea.

The ASEAN Claimants

Given their relative weakness vis-à-vis China, the ASEAN states that are claimants to the Spratly Islands (Malaysia, Brunei, the Philippines, and Vietnam) have the strongest incentive to make use of commercial satellite imagery to monitor developments in the South China Sea. The use of commercial satellite imagery to focus regional and international attention on potentially provocative activities might offer these states some much needed additional diplomatic leverage in dealing with China. Engaging China through regional confidence-building measures is also consistent with the interest of many ASEAN countries in maintaining productive diplomatic and economic relations with Beijing while at the same time resisting any efforts by China to intimidate them by using its military might.[2] The ASEAN countries might even turn to the ARF, which includes China as a member, in developing a formal structure for supporting a cooperative monitoring regime for the South China Sea.

China

Beijing's interests are ostensibly contrary to supporting a cooperative monitoring regime in the region, particularly one focused on the disputed Spratly Islands. China has made expansive sovereignty claims over most of the South China Sea and has even shown a willingness to back up its declarations by periodically using military force. Beyond agreeing to the general principles of peaceful cooperation, Chinese officials tend to avoid international dialogues on the competing claims for the Spratly Islands and serious consideration of various conflict avoidance measures or cooperative development schemes for the Spratlys.

Nonetheless, several reasons exist for China to consider participating in a cooperative monitoring regime. First, if Beijing is unable to discourage a move toward cooperative monitoring for the South China Sea, it could be reluctant to be left out of a regional activity involving territories and waters that it clearly claims. Second, Chinese participation might also be encouraged if Beijing views the proposed regime as a confidence-building measure that is distinct from diplomatic efforts to resolve the competing sovereignty claims over the Spratly Islands. Finally, China's scientific establishment could provide an internal constituency for exploiting the opportunity to improve the nation's remote sensing capabilities through sharing imagery data and technical expertise with the other countries and foreign firms participating in the cooperative monitoring regime. China's engagement in a South China Sea monsoon experiment with the other claimants to the Spratly Islands suggests that opportunities for cooperative monitoring of the South China Sea exist when these scientific activities are divorced from politically sensitive sovereignty issues.[3]

Taiwan

Taiwan represents an anomalous case. Despite the comparable positions that Beijing and Taipei take in asserting sovereign rights to the South China Sea, it is unlikely that both entities would participate together in any formal arrangements for cooperative monitoring regime. Taiwan's role is likely to be limited to tacit cooperation with any informal process or formal structure. Nonetheless, Taiwan has already attempted to use civilian satellite imagery in the past to monitor developments in the Spratly Islands despite the low resolution offered by earlier systems.[4] And given that Taiwan is acquiring its own high-resolution imaging satellite, it will be in a good position to either produce its own satellite imagery, or it can always buy commercial satellite imagery from foreign sources. Taiwan could develop either a formal or informal means for sharing its satellite imagery data with other countries for monitoring developments in the South China Sea.

Facilitators

Receiving active support from various "facilitators"—other countries, businesses, and NGOs concerned with developments in the South China Sea—could be a crucial factor in creating and sustaining a cooperative monitoring regime. Within the region, countries such as Australia, Singapore, and Thailand possess substantial capabilities and expertise in civilian remote sensing, and they have plans to acquire and operate their observation satellites. Thus, they can play an important role in supporting other countries by assisting with the purchase of commercial satellite imagery, as well as by processing and interpreting the imagery data.

Regional organizations are another important resource for supporting a cooperative monitoring regime. ASEAN, with its regional forum for discussing security issues, could play a major role by helping to identify the monitoring and financial requirements for a cooperative monitoring regime focused on both the disputed Spratlys and the broader common interests in the South China Sea. ARF appears to be the most appropriate organization to provide a focal point for developing an agreed-upon monitoring strategy for the South China Sea, as well as hosting the necessary imagery analysis staff.

Another relevant regional institution is the Economic and Social Commission for Asia and the Pacific (ESCAP), which fosters cooperation among Asian countries, including a long-standing program that focuses on remote sensing and geographic information systems (GIS), and has its headquarters in Bangkok.[5] The ESCAP program has already established the basic technical infrastructure and network among the region's remote sensing organizations that are essential for supporting a cooperative monitoring regime focused on the common remote sensing problems of the South China Sea.

Countries, corporations, and nongovernmental organizations (NGOs) outside of the Asia-Pacific region could also make important contributions. The most capable civilian and commercial observation satellites currently belong to the United States, France, Canada, India, and Russia (or their commercial entities). These governments can offer technical assistance and training, as well as encourage their commercial remote sensing firms to adopt data policies that are conducive to a cooperative monitoring regime, which places a premium on imagery data sharing. This is particularly important given that commercial data licensing policies probably need to be relaxed in order to allow the widespread distribution of satellite imagery data among the participants in a South China Sea cooperative monitoring regime.

The United States, Canada, and France can indirectly bolster the efficacy of cooperative monitoring by scheduling their civilian remote sensing satellites to collect images of the Spratly Islands and their surrounding waters on a routine basis. This will ensure that their imagery archives contain a baseline of imagery data that is a prerequisite for providing analysts with the historical imagery necessary for change detection analysis (that is, precise comparisons of old and new imagery scenes to highlight whether or not significant changes have occurred) of Spratly Islands locations. Finally, NGOs can contribute special expertise or even the basic resources that are necessary for developing and sustaining a cooperative monitoring regime for the South China Sea, particularly given the financial constraints on some Asia Pacific governments.

Thus, a cooperative monitoring regime focused on the Spratly Islands disputes is no panacea for eliminating regional conflicts and disagreements. Greater transparency resulting from the use of commercial and civilian observation satellites might even increase the level of tensions in the short run, particularly by denying national governments their ability to downplay politically sensitive developments occurring among the contentious locations in the Spratly Islands. However, this relatively minor discomfiture for government officials must be weighed against the lingering risks of failing to take action to preclude or mitigate a regional conflict in the South China Sea, which, as we have shown, has the potential to serve as a flashpoint with much wider ramifications for international security and stability.

NOTES

1. Commercial remote sensing satellite firms have a strong business interest in not taking sides in diplomatic disputes, as well as in protecting their credibility as providers of credible imagery data to their customers. Countries that have doubts about the loyalties of a particular firm can always turn to other commercial firms or civilian remote sensing satellite enterprises as an alternative source of satellite imagery data, particularly because most commercial firms and countries acquire high-resolution imagery satellites and are willing to sell their imagery data on the global marketplace.

2. For a sense of this ambivalence among many ASEAN countries in dealing with China, see Carolina G. Hernandez and Ralph Cossa, *Security Implications of Conflict in the South China Sea: Perspectives from Asia-Pacific*; and William E. Berry, Jr., *Threat Perceptions in the Philippines, Malaysia, and Singapore* (U.S. Air Force Academy, CO: USAF Institute for National Security Studies, September 1997).

3. See Chapter 8, "A Cooperative Monitoring Regime for the South China Sea," for additional details.

4. "Group Uses Satellite Photos to Monitor Spratlys," Foreign Broadcast Information Service (FBIS) translation of Taipei Voice of Free China broadcast, reprinted in *Daily Report: China* (April 17, 1995), FBIS-CHI-95-073, 79.

5. The broad range of ESCAP activities in promoting regional cooperation in space applications and remote sensing are discussed in some detail in its *Space Technology Applications* annual meeting reports.

Appendix: Occupied Locations among the Spratly Islands, circa 2000

Occupying Country	***Feature Name (English)***	***Feature Geo-Location (lat/long) North East***	***Name in Occupying Country Language***
Malaysia	Ardasier Reef	7°37′ 113°56′	Terumbu Ubi
	Investigator Shoal	8°08′ 114°42′	Terumbu Peninjau
	Mariveles Reef	7°59′ 113°54′	Terumbu Montanani
	Swallow Reef	7°23′ 113°48′	Terumbu Layang Layang
Philippines	Commodore Reef	8°21.5′ 115°13.7′	Rizal Reef
	Flat Island	10°49.5′ 115°50′	Patag
	Lankiam Cay	10°43′ 114°32′	Panata
	Loaita Island	10°40′ 114°25.5′	Dagahoy Dugao
	Nanshan Island	10°44′ 115°48.5′	Lawak
	Northeast Cay	11°27.5′ 114°21′	Parola
	Thitu Island	11°03.2′ 114°17′	Pagasa
	West York Island	11°05.5′ 115°01.5′	Likas
PRC	Chigua Reef	9°55′ 114°29′	Dongmen Jiao
	Cuarteron Reef	8°51.5′ 112°50′	Huayang Jiao
	Fiery Cross Reef	9°33′ 112°53′	Yongshu Jiao
	Gaven Reefs	10°12.5′ 114°13′	Nanxun Jiao
	Johnson Reef (S)	9°42.7′ 114°16.8′	Chigua Jiao
	Mischief Reef	9°55′ 115°32′	Meiji Jiao
	Subi Reef	10°55.5′ 114°05′	Zhubi Jiao

Taiwan	Itu Aba Island	10°23′ 114°21.5′	Tai Ping Dao
Vietnam	Allison Reef	8°49′ 114°06′	Bai Toc Tan
	Amboyna Cay	7°53.8′ 112°55′	Dao An Bang
	Barque Canada Reef	8°10′ 113°18′	Chaw Shai/Chan Chai
	Bombay Castle (Rifleman Bank)	7°56′ 111°40′	—
	Central Reef	8°55′ 112°21′	Con Guia/Trong Sa Dong
	Collins Reef (Johnson Reef-N)	9°45′ 114°13.7′	Bai Vung May/Co Lin
	Cornwallis South Reef	8°43′ 114°11′	Da Nui Le/Nut'le
	Discovery Great Reef	10°04′ 113°51′	Da Lon
	East Reef	8°49′ 112°36′	Con Dong/Da Dong
	Grainger Bank	7°47′ 110°28′	Bai Que Duong
	Kingston Shoal*	7°33′ 111°35′	—
	Ladd Reef	8°37′ 111°40′	Da Lat
	Len Dao	9°46′ 116°43.8′	—
	Namyit Island	10°11′ 114°22′	Nam Yit/Nam Yet
	Orleana Shoal*	70°42′ 114°22′	—
	Pearson Reef	8°57′ 113°40.5′	Hon Sip/Hon Sap
	Pentley Reef	10°24.5′ 114°35′	Do Thi
	Pigeon Reef	8°51.8′ 114°39.2′	Da Tien Nu
	Prince Consort Bank	7°53′ 110°00′	Bai Phuc Nguyen
	Sand Cay	10°22.7′ 114°28.7′	Da Son Ca
	Sin Cowe Island	9°53.2′ 114°19.7′	Dao Sinh Ton/Gac Ma
	Sin Cowe East Island	9°54.8′ 114°33.5′	Sinh Ton Dong
	South Reef	11°23.2′ 114°17.9′	Da Nam
	Southwest Cay	11°25.8′ 114°19.7′	Dao Song Tu Tay
	Spratly Island	8°38.5′ 111°55′	Dao Truong Sa
	Vanguard Bank	7°31.7′ 109°43.7′	Bai Tu Chinch
	West Reef	8°51′ 112°12′	Con Tay/Da Day

*Vietnam reportedly sent troops to occupy two additional submerged reefs in the Spratly Islands (probably Orleana Shoal and Kingston Shoal on Rifleman Bank) in September 1998, but there has been no firm information on these actions. In comparison, Malaysia's move to construct new facilities at Investigator Shoal in June 1999 has been well documented.

Sources: Dan J. Dzurek, *The Spratly Islands Dispute: Who's on First?*, Maritime Briefing 2:1 (Durham, U.K.: International Boundaries Research Unit [IBRU], University of Durham, 1996), Table 2: 56–57; and Rigoberto Tiglao, "Seaside Boom," *Far Eastern Economic Review* (July 8, 1999): 14.

Recommended Readings

THE INFORMATION REVOLUTION, GLOBAL TRANSPARENCY, AND INTERNATIONAL SECURITY

Crocker, Chester A., Fen Osler Hampson, and Pamela Aall, eds. *Herding Cats: Multiparty Mediation in a Complex World.* Washington, D.C.: United States Institute of Peace Press, 1999.

Finel, Bernard I., and Kristin M. Lord. "The Surprising Logic of Transparency." *International Studies Quarterly* 43, June 1999.

Finel, Bernard I., and Kristin M. Lord, eds. *Power and Conflict in the Age of Transparency.* New York: St. Martin's Press, 2000.

Florini, Ann. "The End of Secrecy," *Foreign Policy* 111, Summer 1998, 50–63.

Henry, Ryan, and C. Edward Peartree, eds. *The Information Revolution and International Security*, Washington, D.C.: The CSIS Press, 1998.

Keohane, Robert O., and Joseph S. Nye, Jr. "Power and Interdependence in the Information Age." *Foreign Affairs* 77, September/October 1998, 81–94.

Krepon, Michael, et al. *Global Confidence Building: New Tools for Troubled Regimes.* New York: St. Martin's Press, 1999.

Krepon, Michael, ed. *A Handbook of Confidence-building Measures for Regional Security.* Washington, D.C.: The Henry L. Stimson Center, Handbook No. 1, 1995.

Krepon, Michael, and Amy Smithson, eds. *Open Skies, Arms Control, and Cooperative Security.* Washington, D.C.: The Henry L. Stimson Center, 1992.

Nye, Joseph S. Jr., and William A. Owens. "America's Information Edge." *Foreign Affairs* 75:2, March–April 1996, 20–36.

Rosenau, James N. *States, Sovereignty, and Diplomacy in the Information Age.* Washington, D.C.: United States Institute of Peace, Virtual Diplomacy Series No. 5, February 2000.

Sisk, Timothy D., ed. *New Approaches to International Negotiation and Mediation: Findings from USIP-Sponsored Research.* Washington, D.C.: United States Institute of Peace, Peaceworks No. 30, August 1999.

Solomon, Richard H. "The Global Information Revolution and International Conflict Management," an overview presentation at the 'Virtual Diplomacy' conference, Washington, D.C.: United States Institute of Peace, April 1997.

Wriston, Walter B. "Bits, Bytes, and Diplomacy." *Foreign Affairs* 76:5, September–October 1997, 172–182.

Wriston, Walter B. *The Twilight of Sovereignty: How the Information Revolution Is Transforming Our World.* New York: Scribner's, 1992.

GENERAL ASIA-PACIFIC DEVELOPMENTS

Acharya, Amitav. *A New Regional Order in Southeast Asia: ASEAN in the Post-Cold War Era.* London: International Institute for Strategic Studies, 1993.

Allen, Kenneth W. "China's Approach to Confidence-Building Measures." In Ranjeet K. Singh, ed. *Investigating Confidence-Building Measures in the Asia-Pacific Region.* Washington, D.C.: The Henry L. Stimson Center, May 1999, 53–109.

Ball, Desmond. "A New Era in Confidence Building: The Second-Track Process in the Asia-Pacific Region." *Security Dialogue* 25, June 1994, 157–177.

Bateman, Sam, ed. *Maritime Cooperation in the Asia-Pacific Region: Current Situation and Prospects.* Canberra: Strategic and Defence Studies Centre, Research School of Pacific and Asian Studies, Australian National University, 1999.

Beckman, Robert, Carl Grundy-Warr, and Vivian Forbes. "Acts of Piracy in the Malacca and Singapore Straits." *International Boundaries Research Unit (IBRU) Maritime Briefing* 1:4, 1994.

Bessho, Koro. *Identities and Security in East Asia.* London: International Institute for Strategic Studies, 1999.

Binnendijk, Hans A., and Patrick L. Clawson, eds. *Strategic Assessment 1997: Flashpoints and Force Structure.* Washington, D.C.: U.S. Government Printing Office, 1997.

Blackwill, Robert D., and Paul Dibb, eds. *America's Asian Alliances.* Cambridge, MA: The MIT Press, 2000.

Breckon, Lyall, and Thomas J. Hirschfield. *The Dynamics of Security in the Asia-Pacific Region.* Alexandria, VA: Center for Naval Analyses, January 1996.

Calder, Kent E. "Asia's Empty Tank," *Foreign Affairs* 75, March–April 1996, 55–69.

Carpenter, William M., and David G. Wiencek, eds. *Asian Security Handbook: An Assessment of Political-Security Issues in the Asia-Pacific Region.* Armonk, NY: M.E. Sharpe, 1996.

Carpenter, William M., and David G. Wiencek, eds. *Asian Security Handbook 2000.* Armonk, NY: M.E. Sharpe, 2000.

Chalk, Peter. "The Dynamics of Contemporary Maritime Piracy in Southeast Asia." *Studies in Conflict and Terrorism,* 21:1, 1997.

Chalk, Peter. *Grey Area Phenomena in Southeast Asia: Piracy, Drug Trafficking and Political Terrorism.* Canberra: Strategic and Defense Studies Centre, 1997.

Chalk, Peter. *Non-Military Security and Global Disorder: The Impact of Extremism, Violence and Chaos on National and International Security.* London: Macmillan, 2000.

Chalmers, Malcolm. "Openness and Security Policy in South-east Asia." *Survival* 38:3, Autumn 1996, 83–98.

Clad, James, and Aurora Medina Siy. "The Emergence of Ecological Issues in Southeast Asia." In David Wurfel and Bruce Burton, eds., *Southeast Asia in the New World Order: The Political Economy of a Dynamic Region.* New York: St. Martin's Press, 1996, 52–73.

Cole, Bernard D. *The Great Wall at Sea: China's Navy Enters the Twenty-First Century.* Annapolis, MD: Naval Institute Press, 2001.

Cossa, Ralph A., ed. *Asia Pacific Confidence and Security Building Measures.* Washington, D.C.: The Center for Strategic and International Studies, 1995.

Cossa, Ralph A., ed. "Asia-Pacific Confidence-Building Measures for Regional Security." In Michael Krepon, et. al., *Global Confidence Building: New Tools for Troubled Regimes.* New York: St. Martin's Press, 1999, 17–41.

Dupont, Alan. *The Environment and Security in Pacific Asia.* Adelphi Papers, London: International Institute for Strategic Studies, 1998.

Fisher, Richard D. Jr. "The Accelerating Modernization of China's Military." In Kim R. Holmes and James J. Przystup, eds., *Between Diplomacy and Detente: Strategies for U.S. Relations with China.* Washington, D.C.: The Heritage Foundation, 1997, 97–140.

Hawksley, Humphrey, and Simon Holberton. *Dragon Strike: The Millennium War.* London: Pan Books, 1997. (Fictional account of war in the Spratly Islands and beyond.)

Hitt, Jack. "Bandits in the Global Shipping Lanes." *The New York Times Magazine,* August 20, 2000.

Hung-mao Tien, and Tun-jen Cheng, eds. *The Security Environment in the Asia-Pacific*. Armonk, N.Y.: M.E. Sharpe, 2000.

Hunter, Thomas B., "The Growing Threat of Modern Piracy." *U.S. Naval Institute Proceedings* 125:7, July 1999, 72–74.

Huxley, Tim, and Susan Willett. *Arming East Asia*. Adelphi Papers, London: International Institute for Strategic Studies, 1999.

Kenney, Henry J. *An Analysis of Possible Threats to Shipping in Key Southeast Asian Sea Lanes*. Alexandria, VA: Center for Naval Analyses, 1996.

Khalilzad, Zalmay M., et al. *The United States and a Rising China*. Santa Monica, CA: RAND, 1999.

Kreil, Erik A., David Gregory, Paul Hueper, and John Noer. "Energy in the South China Sea." *Geopolitics of Energy* 19:7, July 1997, 2–7.

Leifer, Michael. *The ASEAN Regional Forum*. Adelphi Papers, London: International Institute for Strategic Studies, 1996.

Lilley, James R., and David Shambaugh. *China's Military Faces the Future*. Armonk, N.Y: M.E. Sharpe, and the American Enterprise Institute, 1999.

Lok, Joris Janssen. "ASEAN Navies Extend their Regional Reach." *Jane's Defense Weekly*, November 27, 1996, 25–28.

Mack, Andrew, ed. *A Peaceful Ocean?: Maritime Security in the Pacific in the Post-Cold War Era*. Canberra: Australian National University, 1993.

Manning, Robert A. *The Asian Energy Factor: Myths and Dilemmas of Energy, Security and the Pacific Future*. New York: St. Martin's Press, 2000.

McDevitt, Michael. "Where Is China's Navy Headed?" *U.S. Naval Institute Proceedings* 127:5, May 2001, 58–61.

Meconis, Charles A., and Michael D. Wallace. *East Asian Naval Weapons Acquisition in the 1990s: Causes, Consequences, and Responses*. Westport, CT: Praeger Publishers, 2000.

Noer, John H., with David Gregory. *Chokepoints: Maritime Economic Concerns in Southeast Asia*. Washington, D.C.: National Defense University Press, 1996.

Pugh, Michael. "Piracy and Armed Robbery at Sea: Problems and Remedies." *Low Intensity Conflict and Law Enforcement* 2:1, 1993.

Renwick, Neil, and Jason Abbott. "Piratical Violence and Maritime Security in Southeast Asia." *Security Dialogue* 30:2, 1999.

Salameh, Mamdouh G. "China, Oil, and the Risk of Regional Conflict." *Survival* 37:4, Winter 1995–1996, 133–146.

Segal, Gerald. "East Asia and the 'Constrainment' of China." *International Security* 20:4, Spring 1996, 107–135.

Simon, Sheldon W. *The Economic Crisis and ASEAN States' Security*. Carlisle Barracks, PA: Strategic Studies Institute, U.S. Army War College, 1998.

Solomon, Richard H., ed. *Asian Security in the 1980s: Problems and Policies for a Time of Transition*. Cambridge, MA: Oelgeschlager, Gunn and Hain, Publishers, 1980.

U.S. Department of Defense. *The United States Security Strategy for the East Asia-Pacific Region*. Washington, D.C.: Office of International Security Affairs, Department of Defense, November 1998.

U.S. Energy Information Administration. "The South China Sea Region." Web site: http://www.eia.doe.gov/emeu/cabs/schina.html (updated annually).

Wanandi, Jusuf. "ASEAN's China Strategy: Towards Deeper Engagement." *Survival* 38:3, Autumn 1996, 117–128.

SOUTH CHINA SEA AND THE SPRATLY ISLANDS DISPUTES

Austin, Greag. "China's Ocean Limit—Time to Settle." A Discussion Paper for the United States Institute of Peace, September 4, 1994.

Baker, John C. "Conflict Potential of the South China Sea Disputes." In William M. Carpenter and David G. Wiencek, eds., *Asian Security Handbook 2000*. Armonk, NY: M.E. Sharpe, 2000, 106–117.

Baker, John C., and David G. Wiencek. "Sat-images Could Be Spratlys Salvation." *Jane's Intelligence Review* 11:2, February 1999, 50–54.

Baker, John C., and David G. Wiencek. "Cooperative Satellite Monitoring for the South China Sea." *Integrated Coastal Zone Management* 2, 2000, 99–104.

Catley, Bob, and Makmur Keliat. *Spratlys: The Dispute in the South China Sea*. Aldershot, UK: Ashgate Publishing Limited, 1997.

Chang, Felix K. "Beijing's Reach in the South China Sea." *Orbis* 40:3, Summer 1996, 353–374.

Chen, Jie, "China's Spratly Policy: With Special Reference to the Philippines and Malaysia." *Asian Survey* 34:10, October 1994, 893–903.

Cheng, Yi Lin, "Taiwan's South China Sea Policy." *Asian Survey* 37:4, April 1997, 323–339.

Chung, Chien, "Confidence-Building Measures in the South China Sea." In Hung-mao Tien and Tun-jen Cheng, eds. *The Security Environment in the Asia-Pacific*. Armonk, N.Y.: M.E. Sharpe, 2000, 259–305.

Cordner, Lee G. "The Spratly Islands Dispute and the Law of the Sea." *Ocean Development and International Law* 25:1, January–March 1994, 61–74.

Cossa, Ralph A. "Security Implications of Conflict in the South China Sea: Exploring Potential Triggers of Conflict." draft paper, Honolulu: Pacific Forum CSIS, 1998.

Dahlby, Tracy. "South China Sea." *National Geographic* 194:6, December 1998, 2–33.

Djalal, Hasjim. "South China Sea Disputes." In Myron H. Nordquist and John Norton Moore, eds., *Security Flashpoints: Oil, Islands, Sea Access and Military Confrontation*. Kluwer Law International Publications, 1998, 109–134.

Djalal, Hasjim, and Ian Townsend-Gault. "Managing Potential Conflicts in the South China Sea: Informal Diplomacy for Conflict Prevention." In Chester A. Crocker, et al., eds., *Herding Cats: Multiparty Mediation in a Complex World*. Washington, D.C.: United States Institute of Peace Press, 1999, 109–133.

Dobson, William J., and M. Taylor Fravel. "Red Herring Hegemon: China in the South China Sea." *Current History* 96:611, September 1997, 258–263.

Dzurek, Daniel J. *The Spratly Islands Dispute: Who's On First?,* Maritime Briefing 2, Durham, U.K.: International Boundaries Research Unit, Mountjoy Research Centre, University of Durham, 1996.

Evans, Paul M. "Assessing the ARF and CSCAP." In Hung-mao Tien and Tun-jen Cheng, eds., *The Security Environment in the Asia-Pacific*. Armonk, N.Y.: M.E. Sharpe, 2000, 154–172.

Gallagher, Michael G. "China's Illusory Threat to the South China Sea." *International Security* 19:1, Summer 1994, 169–194.

Garver, John. "China's Push through the South China Sea: The Interaction of Bureaucratic and National Interests." *The China Quarterly* 131:132, December 1992, 999–1028.

Gupta, Vipin, and Adam Bernstein. "Keeping an Eye on the Islands: Cooperative Remote Monitoring in the South China Sea." In John C. Baker, Kevin M. O'Connell, and Ray A. Williamson, eds., *Commercial Observation Satellites: At the Leading Edge of Global Transparency*. Santa Monica, CA: RAND and the American Society for Photogrammetry and Remote Sensing, 2001, 327–360.

Hamzah, B. A. "Conflicting Maritime Claims in the South China Sea: The Scope for Resolution." In Ross Babbage and Sam Bateman, eds., *Maritime Change: Issues for Asia*. St. Leonards, New South Wales: Allen & Unwin, 1993, 96–109.

Hernandez, Carolina G., and Ralph Cossa, eds. *Security Implications of Conflict in the South China Sea: Perspectives from Asia-Pacific*. Quezon City, Philippines: Institute for Strategic and Development Studies, Inc., and Pacific Forum, CSIS, 1997.

Hull, Richard E. *The South China Sea: Future Source of Prosperity or Conflict in SouthEast Asia?* Washington, D.C.: National Defense University, Institute for National Strategic Studies, No. 60, February 1996.

Ji, Guoxing. *Maritime Jurisdiction in the Three China Seas*. San Diego, CA: Institute on Global Conflict and Cooperation, University of California, 1995.

Joyner, Christopher, C. "The Spratly Islands Dispute: Rethinking the Interplay of Law, Diplomacy, and Geo-politics in the South China Sea." *The International Journal of Marine and Coastal Law* 13:2, 1998, 193–236.

Joyner, Christopher, C. "The Spratly Islands Dispute: What Role for Normalizing Relations between China and Taiwan?" *New England Law Review*, 32: 3, Spring 1998, 819–852.

Joyner, Christopher, C. "The Spratly Islands Dispute in the South China Sea: Problems, Policies, and Prospects for Diplomatic Accommodation." In Ranjeet K. Singh, ed. *Investigating Confidence-Building Measures in the Asia-Pacific Region*. Washington, D.C.: The Henry L. Stimson Center, May 1999, 53–109.

Joyner, Christopher, C. "The Spratly Islands Dispute in the South China Sea." *Integrated Coastal Zone Management*, 1 (1999), 85–89.

Lam, Peng Er. "Japan and the Spratlys Dispute." *Asian Survey* 36, October 1996, 995–1010.

Lee, Lai To. *China and the South China Sea Dialogues*. Westport, CT: Praeger, 1999.

Li, Hui. "Science Overrides Politics for East Asian Monsoon Study." *Science* 280, April 17, 1998, 373.

Lo, Chi-kin. *China's Policy Towards Territorial Disputes: The Case of the South China Sea Islands*. London: Routledge, 1989.

Nugroho, Anton. "The Dragon Looks South." *U.S. Naval Institute Proceedings* 126:3, March 2000, 74–76.

Pitzl, Jerry. "Mischief on Mischief Reef: Chinese Adventures in the Spratly Islands." In Karen A. Mingst and Katsuhiko Mori, eds., *Teaching International Affairs with Cases: Cross National Perspectives*. Boulder, CO: Westview Press, 1997, 175–180.

Samuels, Marwyn S. *Contest for the South China Sea*. New York: Methuen, 1982.

Schofield, Clive. "A Code of Conduct for the South China Sea?" *Jane's Intelligence Review*, November 2000, 36–40.

Snyder, Craig, "The Implications of Hydrocarbon Development in the South China Sea." *International Journal* 52, Winter 1996–1997, 142–158.

Snyder, Scott. *The South China Sea Dispute: Prospects for Preventive Diplomacy*, A Special Report of the United States Institute of Peace. Washington, D.C.: United States Institute of Peace, August 1996.

Sokolsky, Richard, Angel Rabasa, and C. R. Neu. *The Role of Southeast Asia in U.S. Strategy Toward China*. Santa Monica, CA: RAND, 2000.

Stenseth, Leni. *Nationalism and Foreign Policy—The Case of China's Nansha Rhetoric*. Oslo: University of Oslo, Centre for Development and the Environment, December 1998.

Storey, Ian. "Manila Looks to USA for Help over Spratlys." *Jane's Intelligence Review* 11:8, August 1999,. 46–50.

Studeman, Michael. "Calculating China's Advances in the South China Sea." *Naval War College Review* 60:2, Spring 1998, 68–90.

Thomas, Bradford L., and Daniel J. Dzurek. "The Spratly Islands Dispute." *Geopolitics and International Boundaries* 1:3, Winter 1996, 300–326.

Valencia, Mark J. *China and the South Sea Disputes.* London: International Institute for Strategic Studies, 1995.

Valencia, Mark J. "Energy and Insecurity in Asia." *Survival,* 39: 3, Autumn 1997, 85–106.

Valencia, Mark J. "The Spratly Imbroglio in the Post-Cold War Era." In David Wurfel and Bruce Burton, eds., *Southeast Asia in the New World Order: The Political Economy of a Dynamic Region.* New York: St. Martin's Press, 1996, 244–269.

Valencia, Mark J. "Troubled Waters," *The Bulletin of the Atomic Scientists,* 53: 1, January/February 1997, 49–54.

Van Dyke, Jon M., and Dale L. Bennett. "Islands and the Delimitation of Ocean Space in the South China Sea." In Elisabeth Mann Borgese, Norton Ginsburg, and Joseph R. Morgan, eds., *Ocean Yearbook No. 10,* Chicago: The University of Chicago Press, 1993, 54–89.

Wain, Barry. "Nowhere to Hide." *Asian Wall Street Journal,* April 20, 2000.

Wang, Liyu, and Peter H. Pearse. "The New Legal Regime for China's Territorial Sea." *Ocean Development and International Law* 25:4, October–December 1994, 431–442.

Wright, Robert. "Private Eyes." *The New York Times Magazine,* September 5, 1999.

Wu, Samuel S. G., and Bruce Bueno de Mesquita. "Assessing the Dispute in the South China Sea: A Model of China's Security Decision Making." *International Studies Quarterly* 38:3, September 1994, 379–403.

Zhiguo, Gao. "The South China Sea: From Conflict to Cooperation?" *Ocean Development and International Law* 25, 1994, 345–359.

COMMERCIAL AND CIVILIAN SATELLITE REMOTE SENSING

Abiodun, Adigun Ade. "Remote Sensing in the Information Age." *Space Policy* 14:4, November 1998, 229–238.

Anselmo, Joseph C. "Commercial Imagery Aids Bosnian Peace Mission." *Aviation Week and Space Technology,* February 5, 1996, 60.

Baker, James D. *Planet Earth: The View from Space.* Cambridge, MA: Harvard University Press, 1990.

Baker, John C. *Trading Away Security?: The Clinton Administration's 1994 Decision on Imaging Satellite Exports*. Pew Case Studies in International Affairs, Washington, D.C.: Institute for the Study of Diplomacy, Georgetown University, 1997.

Baker, John C., Kevin M. O'Connell, and Ray A. Williamson, eds. *Commercial Observation Satellites: At the Leading Edge of Global Transparency*. Santa Monica, CA: RAND and the American Society for Photogrammetry and Remote Sensing, 2001.

Bruginoni, Dino. "Aerial Photography: Reading the Past, Revealing the Future." *Smithsonian* 14:12, March 1984, 150–161.

Bruginoni, Dino. "The Art and Science of Photoreconnaissance." *Scientific American* 274: 3, March 1996, 78–85.

Campbell, James B. *Introduction to Remote Sensing*, New York: The Guilford Press, 1996.

Colwell, Robert N., editor-in-chief. *Manual of Remote Sensing I*, Falls Church, VA: American Society of Photogrammetry, 1983.

Day, Dwayne A., John M. Logsdon, and Brian Latell, eds. *Eye in the Sky: The Story of the Corona Spy Satellites*. Washington, D.C.: Smithsonian Institution Press, 1998.

Dehqanzada, Yahya A., and Ann M. Florini. *Secrets for Sale: How Commercial Satellite Imagery Will Change the World*. Washington, D.C.: Carnegie Endowment for International Peace, 1998.

Doyle, Stephen E. *Civil Space Systems: Implications for International Security*, United Nations Institute for Disarmament Research, Dartmouth Publishing Company, 1994.

Ferster, Warren. "Private Spacecraft Imagery Evolves as Treaty Tool." *Space News*, October 20–26, 1997, 3.

Florini, Ann M. "The Opening Skies: Third Party Imaging Satellites and US Security." *International Security* 13:2, Fall 1988, 91–123.

Fuller, D. O. "Satellite Remote Sensing of Biomass Burning Using Optical and Thermal Sensors." *Progress in Physical Geography* 24:4, 2000, 543–561.

Fuller, D. O. "Burned Area in Kalimantan, Indonesia Mapped with NOAA-AVHRR and Landsat TM Imagery." *International Journal of Remote Sensing* 22: 4, March 2001, 691–697.

Gabbard, C. Bryan, Kevin M. O'Connell, George S. Park, and Peter J. E. Stan. *Emerging Markets of the Information Age: A Case Study in Remote Sensing Data and Technology*, Santa Monica, CA: RAND, 1996.

Gaddis, John Lewis. "Learning to Live with Transparency: The Emergence of a Reconnaissance Satellite Regime." In John Lewis Gaddis, *The Long Peace: Inquiries Into the History of the Cold War*, New York: Oxford University Press, 1987, 195–214.

George, Hubert. "Remote Sensing of Earth Resources: Emerging Opportunities for Developing Countries." *Space Policy* 14, 1998, 27–37.

Grundhauser, Larry K. "Sentinels Rising: Commercial High-Resolution Satellite Imagery and Its Implications for US National Security." *Airpower Journal* 12:4, Winter 1998, 61–80.

Gupta, Vipin. "New Satellite Images for Sale." *International Security* 20:1, Summer 1995, 94–125.

Gupta, Vipin, and Frank Pabian. "Investigating the Allegations of Indian Nuclear Test Preparations in the Rajasthan Desert: A CTB Verification Exercise Using Commercial Satellite Imagery." *Science and Global Security* 6, 1996, 101–189.

Harris, Ray. *Satellite Remote Sensing: An Introduction*, London: Routledge & Kegan Paul, 1987.

Jasani, Bhupendra, and Toshibomi Sakata. *Satellites for Arms Control and Crisis Monitoring*. Oxford: Oxford University Press, 1987.

Jensen, John R. *Remote Sensing of the Environment: An Earth Resource Perspective*. Upper Saddle River, N.J.: Prentice Hall, 2000.

Jingxuan, Lu, "Marine Oil Spill Surveillance and Mapping Using Remote Sensing in Singapore." *Integrated Coastal Zone Management* 1, 2000, 217–222.

Krepon, Michael, Peter Zimmerman, Leonard Spector, and Mary Umberger, eds. *Commercial Observation Satellites and International Security*. New York: St. Martin's Press, 1990.

Lillesand, Thomas M., and R. W. Kiefer. *Remote Sensing and Image Interpretation*. New York: Wiley & Sons, 2000.

Litfin, Karen T. "The Status of the Statistical State: Satellites and the Diffusion of Epistemic Sovereignty." *Global Society* 13: 1, 1999, 95–116.

Mack, Pamela E. *Viewing the Earth: The Social Construction of the Landsat Satellite System*, Cambridge, MA: The MIT Press, 1990.

McDonald, Robert A., ed. *Corona between the Sun and the Earth: The First NRO Reconnaissance Eye in Space*. Bethesda, MD: American Society for Photogrammetry and Remote Sensing, 1997.

Petrocchi, Andrea. "Slick Sensing: Monitoring Oil Spills with Satellite Imagery." *Geo Info Systems* 10:5, May 2000, 32–36.

Philipson, Warren R., editor-in-chief. *Manual of Photographic Interpretation*. Bethesda, MD: American Society for Photogrammetry and Remote Sensing, 1997.

Seitz, Patrick. "Singapore Grows as Major Earth Imagery Center." *Space News*, October 28–November 3, 1996, 28.

Steinberg, Gerald M. *Dual Use Aspects of Commercial High-Resolution Imaging Satellites*. Ramat Gan, Israel: The Begin-Sadat Center for Strategic Studies, Bar-Ilan University, February 1998.

Tahu, George J., John C. Baker, and Kevin M. O'Connell. "Expanding Global Access to Civilian and Commercial Remote Sensing Systems and Data: Implications and Policy Issues." *Space Policy* 14, August 1998, 179–188.

Uhlir, Paul F. "From Spacecraft to Statecraft: The Role of Earth Observation Satellites in the Development and Verification of International Environmental Protection Agreements." *GIS Law* 2:3, Winter 1995, 1–15.

U.S. Congress, Office of Technology Assessment. *Civilian Satellite Remote Sensing: A Strategic Approach.* Washington, D.C.: U.S. Government Printing Office, September 1994.

U.S. Congress, Office of Technology Assessment. *Civilian Satellite Systems and Applications.* Washington, D.C.: U.S. Government Printing Office, July 1993.

U.S. Congress, Office of Technology Assessment. *Verification Technologies: Cooperative Aerial Surveillance in International Agreements.* Washington, D.C.: U.S. Government Printing Office, July 1991.

U.S. Department of Commerce. *Remote Sensing for Coastal Zone Managers: An Overview.* Washington, D.C.: U.S. Department of Commerce, National Oceanic and Atmospheric Administration, National Ocean Service, February 1997.

Wagner, Caroline S. *International Agreements on Cooperation in Remote Sensing and Earth Observation.* Santa Monica, CA: RAND, MR-972-OSTP, 1998.

Williamson, Ray A. "The Landsat Legacy: Remote Sensing Policy and the Development of Commercial Remote Sensing." *PE&RS: Photogrammetric Engineering and Remote Sensing* 63:7, July 1997, 877–885.

Index

About the Contributors

JOHN C. BAKER is a Technology Policy Analyst at RAND, Arlington, Virginia. His work is concerned with the international security implications of commercial observation satellites and other dual-use space technologies. He is the co-editor (with Kevin M. O'Connell and Ray A. Williamson) of *Commercial Observation Satellites: At the Leading Edge of Global Transparency* (2001). Previously, as a senior staff member at the Space Policy Institute in the Elliott School of International Affairs at the George Washington University, he directed the Institute's South China Sea Remote Sensing project. He has also been a research associate at the International Institute for Strategic Studies (London), the Pacific-Sierra Research Corporation, and the Brookings Institution.

ADAM BERNSTEIN is a physicist working in the Sensor Development Department at the Sandia National Laboratories office in Livermore, California. He is a specialist in the development and use of technologies for treaty verification and cooperative monitoring, with an emphasis on nuclear arms control. Bernstein contributed to the imagery analysis for the South China Sea Remote Sensing project that was the basis for this book. He is a coauthor (with Vipin Gupta) of "Keeping an Eye on the Islands: Cooperative Remote Monitoring in the South China Sea" in *Commercial Observation Satellites: At the Leading Edge of Global Transparency* (2001). He has also written *Monitoring Large Enrichment Plants Using Thermal Imagery from Commercial Satellites: A Case Study* (2000).

PETER CHALK is a Policy Analyst working in the Project Air Force and National Defense Research Institute divisions at RAND in Arlington, Virginia. He has recently worked on projects examining unconventional security threats in Southeast Asia, new strategic challenges for the U.S. Air Force (USAF) in Latin America, and evolving trends in national and international terrorism. He also was a major contributor to the First Annual Report of the Gilmore Commission, a congressionally mandated advisory panel established to assess U.S. domestic response capabilities for terrorism involving weapons of mass destruction. Prior to joining RAND, Chalk was an Assistant Professor of Politics at the University of Queensland, Brisbane, and a Postdoctoral Fellow in the Strategic and Defense Studies Centre of the Australian National University, Canberra. In addition to his academic posts, Chalk has acted as a research consultant in the United Kingdom, Canada, and Australia. He is author of *Non-Military Security and Global Disorder: The Impact of Extremism, Violence and Chaos on National and International Security* (2000) and *West European Terrorism and Counter-Terrorism: The Evolving Dynamic* (1996), in addition to numerous journal articles and other scholarly publications in the security studies field. He is a contributing editor to *Studies in Conflict and Terrorism* and represents the United States on the Council for Security Cooperation in the Asia Pacific's (CSCAP) Working Group on Transnational Crime.

DOUGLAS O. FULLER joined the faculty of the Department of Geography at The George Washington University in Washington, D.C., as Assistant Professor in 1997 after working in Senegal, West Africa, as an ecologist for the United Nations. Prior to working in Senegal, Fuller completed his Ph.D. studies in biogeography and remote sensing at the University of Maryland, College Park, and held a NASA Global Change Fellowship (1992–1993). During that time, he worked as a research associate at the NASA Goddard Space Flight Center in Greenbelt, Maryland, on a time-series analysis of satellite imagery. Before his involvement in remote sensing, Fuller held several positions in environmental conservation in Washington, D.C., at the World Wildlife Fund and The World Resources Institute.

VIPIN GUPTA is a remote sensing and arms control specialist at the Sandia National Laboratories office in Livermore, California, as well as a senior technical staff member in Sandia's Systems Research Department. He is a pioneer in applying civilian and commercial satellite imagery to international security problems. Gupta led the aerial and satellite imagery analysis for the South China Sea Remote Sensing project, which is the basis for this book. He is a coauthor (with Adam Bernstein) of "Keeping an Eye on the Islands: Cooperative Remote Monitoring in the South

China Sea" in *Commercial Observation Satellites: At the Leading Edge of Global Transparency* (2001). Gupta has published and presented extensively on the broader security implications of commercial and civilian observation satellites. His articles include "New Satellite Images for Sale," *International Security* 20 (Summer 1995), and an in-depth imagery analysis of possible nuclear weapon test preparations by India in 1995 (with Frank Pabian) in "Investigating the Allegations of Indian Nuclear Test Preparations in the Rajasthan Desert: A CTB Verification Exercise Using Commercial Satellite Imagery," *Science and Global Security* 6 (1996).

CHRISTOPHER C. JOYNER is a Professor of International Law in the Department of Government at Georgetown University, where he teaches courses on international law and organizations, U.S. foreign policy, and global environmental law. From 1989–1994, he directed the American Society of International Law's Project on the United Nations and the International Legal Order, and was the co-editor of *United Nations Legal Order* (1995) and the editor of *The United Nations and International Law* (1997). A member of the Governing Council of the Academic Council on the United Nations (ACUNS) and the International Law Association's Committee on the Law of the Sea, he has served twice on the ASIL Executive Council and three times past Chair of the International Law Section of the International Studies Association. Professor Joyner has written more than 300 articles in law reviews and published twelve books, among them *Antarctica and the Law of the Sea* (1992) and *Governing the Frozen Commons: The Antarctic Regime and Environmental Protection* (1998), and a forthcoming publication, *International Law in the 21st Century: Legal Rules for Global Governance.*

ERIK KREIL is an international energy analyst for the U.S. Energy Information Administration (EIA), and covers energy security issues and world oil markets for EIA. He has assisted the National Defense University and U.S. Army War College in conducting their strategic evaluations of South China Sea issues, and has supported the Department of Energy's work with the U.S. Naval War College in this region.

BRADFORD L. THOMAS, until his untimely death in 1999, was an International Boundary Consultant in Washington, D.C., and a Professorial Lecturer in the Department of Geography at The George Washington University, where he taught remote sensing and air photo interpretation. He was a recognized expert on the Spratly Islands dispute and published several articles on this subject including, "The Spratly Islands Dispute," in *Geopolitics and International Boundaries 1* (Winter 1996), with Daniel J. Dzurek. From 1986–1997, Thomas worked for the Office of The

Geographer and Global Issues, U.S. Department of State, serving from 1991 as Chief of the Cartography and Boundary Analysis Division. From 1981–1986, he worked for the Office of the Coordinator for Maps and Publications at the State Department. His earlier positions included Director of Cartographic Services Laboratory at the University of Maryland, and faculty positions in geography at Kent State University, San Francisco State College, and the University of Minnesota.

DAVID G. WIENCEK heads the consulting company, International Security Group, Inc., which specializes in international political risk assessments, East and Southeast Asian political-security affairs, weapons of mass destruction and proliferation studies, and terrorism issues. He previously served as the Director of Research and Information Services at Armor Group/Parvus International Corporation, an international security and risk management firm in Washington, D.C. His articles have appeared in *The Washington Times*, *Jane's Intelligence Review*, *Defense News*, and *Global Defence Review*. He is also the co-editor with William M. Carpenter of *Asian Security Handbook 2000* (2000).